Nonlinear Adiabatic Evolution of Quantum Systems

Jie Liu · Sheng-Chang Li
Li-Bin Fu · Di-Fa Ye

Nonlinear Adiabatic Evolution of Quantum Systems

Geometric Phase and Virtual Magnetic Monopole

Springer

Jie Liu
Institute of Applied Physics
and Computational Mathematics
Beijing, China

Sheng-Chang Li
Xi'an Jiaotong University
Xi'an, China

Li-Bin Fu
China Academy of Engineering Physics
Beijing, China

Di-Fa Ye
Institute of Applied Physics
and Computational Mathematics
Beijing, China

ISBN 978-981-13-2642-4 ISBN 978-981-13-2643-1 (eBook)
https://doi.org/10.1007/978-981-13-2643-1

Library of Congress Control Number: 2018955156

This Springer imprint is published by the registered company Springer Nature Singapore Pte Ltd.
The registered company address is: 152 Beach Road, #21-01/04 Gateway East, Singapore 189721, Singapore

Preface

Adiabatic theory of both classical and quantum systems plays an important role in addressing various problems with multi-time-scale characteristics, ranging from atomic and molecular processes to the evolution of the universe. In the classical case, the well-known adiabatic theorem states, in terms of action-angle variables, that the action is the adiabatic invariant and that if the Hamiltonian is taken around a given cycle in parameter space, then the angle variable conjugate to the action acquires a purely geometrical quantity, which is termed the Hannay angle. The adiabatic theorem of quantum systems, however, becomes much more intricate due to the involvement of the complex-valued wave function/probability amplitude.

A complete theory for the adiabatic evolution of quantum systems rests on three pillars. First, Born and Fock proved the quantum adiabatic theorem shortly after the discovery of the Schrödinger equation. The adiabatic theorem states that the probability on each instantaneous (nondegenerate) eigenstate remains constant when the external condition changes slowly in time. Second, in addition to the typical dynamical phase, given by the time integral of the eigenenergy, the phase of an evolving eigenstate has a geometric part, called the Berry phase, that depends only on the geometric path in the parameter space. Third, this geometric phase can be interpreted as the flux of a virtual magnetic monopole field through the surface enclosed by the closed circuit in the parameter space. The adiabatic theory has played a crucial role in the preparation and control of quantum states. The Berry phase and related geometric phases have important applications in modern physics, such as in high-precision quantum measurement, quantum information processing, quantum computing, and condensed-matter physics.

In this book, we generalize the adiabatic theory to the nonlinear evolution of quantum systems. In physics, the nonlinearity has been introduced as possible modifications of quantum mechanics on the fundamental level. However, our motivation derives mainly from the practical applications of adiabatic control of Bose-Einstein condensates (BECs), which can often be accurately described by the nonlinear Schrödinger equation. Here, the nonlinearity stems from a mean-field treatment of the interactions between coherent atoms. The appearance of nonlinearity leads not only to the lack of unitarity but also to the absence of the superposition principle. We overcome these challenges by combining ideas from classical adiabatic dynamics and quantum geometric phases. The developed theory of nonlinear quantum adiabatic evolution is expected to be useful in guiding adiabatic manipulation of the condensate atoms and other nonlinear systems.

The book is organized as follows. In Chap. 1, we introduce the basic concepts of adiabatic theory, such as the adiabatic invariant, the Hannay angle, the adiabatic theorem, the Berry phase, and the virtual magnetic monopole. Some typical examples of adiabatic evolution are presented. In Chap. 2, we discuss the physical origins of the nonlinearity in quantum many-body systems. The nonlinear adiabatic theory, including the adiabatic evolution of the quantum states and the nonlinear geometric phase, is introduced. In Chap. 3, we discuss the commutability between the adiabatic limit and the semiclassical limit. We show the relationship between the quantum Berry phase, the classical Hannay angle, and the mean-field geometric phase of an interacting bosonic many-body system. In Chap. 4, we introduce exotic virtual magnetic monopoles and fields such as the disk-shaped virtual magnetic field, fractional virtual magnetic monopole, and virtual magnetic monopole chain. In Chap. 5, we describe selected important applications of nonlinear adiabatic evolution in the geometric phase, in tunneling dynamics, and in quantum interference. We anticipate that readers will find this book useful in providing basic concepts and important applications on nonlinear adiabatic evolution of quantum systems.

I am deeply indebted to my beloved family for their continued support. I am also grateful to my students Li-Da Zhang, Fu-Quan Dou, Hui Cao, Qiang Wang, and Wen-Yuan Wang for reading parts of the manuscript and contributing useful remarks. In particular, I thank Profs. Q. Niu, B. Wu, B. B. Hu, and B. W. Li for long-term fruitful collaborations. Some of our previous collaborating works are included in this book.

Beijing, China
January 2018

Jie Liu

Contents

Chapter 1
Introduction to Adiabatic Evolution

Abstract In this chapter, we introduce the basic concepts of adiabatic theory in both classical and quantum systems. We discuss classical adiabatic motion, introduce the concepts of the classical adiabatic invariant and the Hannay angle, and give three examples: the one-dimensional harmonic oscillator, the celestial two-body problem, and the Foucault pendulum. We describe quantum adiabatic evolution, present the quantum adiabatic theorem, and describe the adiabatic geometric phase (specifically, the Berry phase) and the virtual magnetic monopole. Five examples of quantum adiabatic evolution are shown. We also discuss classical-quantum correspondence.

1.1 Classical Adiabatic Motion

1.1.1 Classical Adiabatic Invariant

We introduce the adiabatic invariant, which is the conserved quantity in adiabatic evolution. For convenience, we consider one-dimensional finite motion of a mechanical system and use the parameter R to describe the properties of the system or of the external field in which it is placed [1]. We assume that the parameter $R(t)$ slowly varies with time because of the external field influence. In other words, the change of the parameter R is very small during one motion period of the system T, i.e.,

$$\frac{dR}{dt} \ll 1. \tag{1.1}$$

Clearly, if the parameter R is time independent, then the energy of the system E is conserved, and the system executes periodic motion. If the parameter R is time dependent, then the energy of the system is not conserved. However, because the parameter changes very slowly with time, the rate of the energy change dE/dt should also be very small. Averaging this rate of change over the motion period and eliminating the fast oscillation part, one can obtain the stable value of dE/dt denoting the slow change of the system energy; this value is proportional to the rate of the parameter dR/dt. In fact, the slowly varying quantity E is a function of the parameter

J. Liu et al., *Nonlinear Adiabatic Evolution of Quantum Systems*,
https://doi.org/10.1007/978-981-13-2643-1_1

R, and the dependence of E on R can be expressed in terms of their combination equaling a constant quantity. The quantity that remains invariant during the evolution of a system with a slowly varying parameter is called the "adiabatic invariant".

The Hamiltonian of the system is $H(p, q; R)$ (here, p and q are a pair of canonical variables corresponding to the generalized momentum and the generalized coordinate, respectively), and the derivative of energy versus time is

$$\frac{dE}{dt} = \frac{\partial H}{\partial R}\frac{dR}{dt}. \tag{1.2}$$

The right-hand side of the equation depends not only on the slow variable R but also on the fast variables p and q. To find the stable variation rule for the system energy, one must apply averaging to (1.2) over the entire motion period T. The parameter R changes very slowly; therefore, the change of dR/dt is also slow. As a result, one can bring dR/dt out of the averaging operation, i.e.,

$$\overline{\frac{dE}{dt}} = \overline{\frac{\partial H}{\partial R}}\frac{dR}{dt}. \tag{1.3}$$

Note that when one applies averaging to the function $\partial H/\partial R$, one considers only p and q as variables. In other words, this averaging operation is used when the parameter R remains constant. In explicit form, one has

$$\overline{\frac{\partial H}{\partial R}} = \frac{1}{T}\int_0^T \frac{\partial H}{\partial R}dt. \tag{1.4}$$

From the Hamilton equation $dq/dt = \partial H/\partial p$, one has

$$dt = \frac{dq}{\partial H/\partial p}. \tag{1.5}$$

Applying this equation, the integration (1.4) with respect to the time t can be replaced with integration with respect to the generalized coordinate q. Furthermore, one can rewrite the motion period T in the following integral form:

$$T = \int_0^T dt = \oint \frac{dq}{\partial H/\partial p}, \tag{1.6}$$

where $\oint$ denotes the integral over the whole region of the change of the generalized coordinate during one period of motion. (For rotation, the coordinate q becomes the rotation angle ϕ, and the integral is over a cycle, i.e., from 0 to 2π). Based on this transformation, the Eq. (1.3) can be rewritten as

$$\overline{\frac{dE}{dt}} = \frac{dR}{dt}\frac{\oint \frac{\partial H/\partial R}{\partial H/\partial p}dq}{\oint \frac{1}{\partial H/\partial p}dq}. \tag{1.7}$$

Note that the integral in this equation is applied along the trajectory of motion for a fixed parameter R. Clearly, when the motion follows such a trajectory, the Hamiltonian retains a constant E. As a result, the generalized momentum p can be expressed as a given function of the generalized coordinate q and the independent parameters E and R, i.e., $p(q;\, E, R)$. Then, by computing the derivative of $H(p, q;\, R) = E$ versus the parameter R, one has

$$\frac{\partial H}{\partial R} + \frac{\partial H}{\partial p}\frac{\partial p}{\partial R} = \frac{dE}{dR} = 0, \tag{1.8}$$

i.e.,

$$\frac{\partial H/\partial R}{\partial H/\partial p} = -\frac{\partial p}{\partial R}. \tag{1.9}$$

Substituting this equation back into the integral in the numerator on the right-hand side of Eq. (1.7) and expressing the core of the denominator as $\partial p(q;\, E, R)/\partial E$, one obtains

$$\overline{\frac{dE}{dt}} = -\frac{dR}{dt}\frac{\oint \frac{\partial p}{\partial R} dq}{\oint \frac{\partial p}{\partial E} dq}, \tag{1.10}$$

i.e.,

$$\oint \left(\frac{\partial p}{\partial E}\overline{\frac{dE}{dt}} + \frac{\partial p}{\partial R}\frac{dR}{dt} \right) dq = 0. \tag{1.11}$$

One introduces the integral along the motion trajectory for the given parameters E and R:

$$I = \frac{1}{2\pi}\oint p dq. \tag{1.12}$$

Equation (1.11) can be expressed as

$$\overline{\frac{dI}{dt}} = 0. \tag{1.13}$$

This result implies that if the parameter R slowly varies with time, then I remains constant, i.e., I is the adiabatic invariant of the system.

We now employ the concept of phase space to show the geometric meaning of the integral in Eq. (1.12). For a system with one degree of freedom, the phase space simplifies to the phase plane spanned by the generalized coordinate p and the generalized momentum q, and the phase trajectory for the periodic motion is a closed orbit in this phase plane. The integral (1.12) along this orbit gives the area enclosed by the closed trajectory. Thus, the adiabatic invariant can be expressed in the following integral form:

$$I = \frac{1}{2\pi} \int \int dp dq. \tag{1.14}$$

We now discuss a near-integrable Hamiltonian for small perturbations and for slow (or "adiabatic") perturbations [2]. For small perturbations, the Hamiltonian has the general form

$$H = H_0(\boldsymbol{I}, t) + \varepsilon H_1(\boldsymbol{I}, \boldsymbol{\theta}, t) + \cdots , \tag{1.15}$$

where H_0 describes completely integrable motion, $\boldsymbol{I}$ and $\boldsymbol{\theta}$ are the N-dimensional actions and angles, and ε is a small parameter characterizing the magnitude of the nonintegrable part of H. For small perturbations, the derivatives of H_0 and H_1 are assumed to be of the same order as H_0 and H_1 themselves, i.e.,

$$\left| \frac{\partial H_0}{\partial t} \right| \sim |H_0|, \quad \left| \frac{\partial H_1}{\partial t} \right| \sim |H_1|, \quad \text{etc.} \tag{1.16}$$

For slow perturbations, the terms produced by differentiation are assumed to be smaller by order ε than the terms from which they are derived, e.g., for slow time variation,

$$\left| \frac{\partial H_0}{\partial t} \right| \sim \varepsilon |H_0|, \quad \text{etc.} \tag{1.17}$$

To keep track of this ordering, one can often insert the small parameter ε and write

$$H_0 = H_0(\varepsilon t) \tag{1.18}$$

such that

$$\frac{\partial H_0}{\partial t} = \varepsilon H_0', \tag{1.19}$$

where the prime notation denotes differentiation with respect to the argument $\tau = \varepsilon t$.

In this section, we are interested in systems for which the variation in all but one of the degrees of freedom, as well as in time, is slow [2]. Accordingly, one can write the Hamiltonian in the form

$$H = H_0(I, \varepsilon \boldsymbol{\eta}, \varepsilon t) + \varepsilon H_1(I, \theta, \varepsilon \boldsymbol{\eta}, \varepsilon t) + \cdots , \tag{1.20}$$

where I and θ are the action-angle variables for the unperturbed ($\varepsilon = 0$) motion in the single fast degree of freedom and $\boldsymbol{\eta} = (\boldsymbol{p}, \boldsymbol{q})$ are the "slow" canonical variables, not necessarily in action-angle form, for the remaining degrees of freedom. Since the system is effectively one-dimensional when $\varepsilon = 0$, this system is integrable, and I and θ can always be found. The small parameter ε in (1.20) "automatically" keeps track of the ordering when one differentiates H to construct the perturbation series; this parameter can be set to unity at the end of the calculation.

One can construct to first order the classical adiabatic invariant for the Hamiltonian (1.20). In zero order, the invariant is the action I associated with the fast degree of freedom. To calculate the effect of the perturbation εH_1, one can find a transformation from $(I, \theta, \boldsymbol{\eta})$ to $(\bar{I}, \bar{\theta}, \bar{\boldsymbol{\eta}})$ such that the new Hamiltonian

$$\bar{H} = \bar{H}_0 + \varepsilon \bar{H}_1 + \cdots \tag{1.21}$$

is independent of the "fast" phase variable $\bar{\theta}$. Introducing the near-identity generating function

$$S = \bar{I}\theta + \bar{\boldsymbol{p}} \cdot \boldsymbol{q} + \varepsilon S_1(\bar{I}, \theta, \bar{\boldsymbol{p}}, \boldsymbol{q}, t) + \cdots , \tag{1.22}$$

one has, to first order, the transformations

$$I = \bar{I} + \varepsilon \frac{\partial S_1}{\partial \bar{\theta}}, \tag{1.23}$$

$$\theta = \bar{\theta} - \varepsilon \frac{\partial S_1}{\partial \bar{I}}, \tag{1.24}$$

$$\boldsymbol{p} = \bar{\boldsymbol{p}} + \varepsilon \frac{\partial S_1}{\partial \bar{\boldsymbol{q}}}, \tag{1.25}$$

$$\boldsymbol{q} = \bar{\boldsymbol{q}} - \varepsilon \frac{\partial S_1}{\partial \bar{\boldsymbol{p}}}. \tag{1.26}$$

Inserting these into H_0 and expanding to first order in ε, one has

$$H_0(I, \varepsilon\boldsymbol{\eta}, \varepsilon t) = H_0(\bar{I}, \varepsilon\bar{\boldsymbol{\eta}}, \varepsilon t) + \varepsilon\omega \frac{\partial S_1}{\partial \bar{\theta}}, \tag{1.27}$$

where $\omega = \partial H_0 / \partial \bar{I}$ is the fast frequency. Note that the terms in

$$-\frac{\partial H_0}{\partial \bar{\boldsymbol{q}}} \cdot \frac{\partial S_1}{\partial \bar{\boldsymbol{p}}}, \quad \frac{\partial H_0}{\partial \bar{\boldsymbol{p}}} \cdot \frac{\partial S_1}{\partial \bar{\boldsymbol{q}}} \tag{1.28}$$

are second order in ε and can be neglected. The canonical transformation equation is

$$\bar{H}(\bar{I}, \bar{\theta}, \varepsilon\bar{\boldsymbol{\eta}}, \varepsilon t) = H(I, \theta, \varepsilon\boldsymbol{\eta}, \varepsilon t) + \varepsilon \frac{\partial S(\bar{I}, \theta, \varepsilon\bar{\boldsymbol{p}}, \varepsilon\boldsymbol{q}, \varepsilon t)}{\partial(\varepsilon t)}. \tag{1.29}$$

Expanding $\bar{H}$, H, and S using the above transformations and equating like powers of ε, to zero order, one obtains

$$\bar{H}_0(\bar{I}, \varepsilon\bar{\boldsymbol{\eta}}, \varepsilon t) = H_0(\bar{I}, \varepsilon\bar{\boldsymbol{\eta}}, \varepsilon t), \tag{1.30}$$

and to first order, one has

$$\bar{H}_1(\bar{I}, \bar{\theta}, \varepsilon\bar{\boldsymbol{\eta}}, \varepsilon t) = \omega \frac{\partial S_1}{\partial \bar{\theta}} + H_1(\bar{I}, \bar{\theta}, \varepsilon\bar{\boldsymbol{\eta}}, \varepsilon t), \tag{1.31}$$

where $S_1 = S_1(\bar{I}, \bar{\theta}, \varepsilon\bar{\eta}, \varepsilon t)$. Again, the term $\partial S_1/\partial t$ in (1.29) is second order and has been omitted from (1.31).

To make $\bar{H}_1$ independent of $\bar{\theta}$, one can choose S_1 to eliminate the oscillating part (in $\bar{\theta}$) of H_1. Holding the slow angle variables fixed, one can define the average over $\bar{\theta}$ alone as

$$\langle \bar{H}_1 \rangle_{\bar{\theta}} = \frac{1}{2\pi} \int_0^{2\pi} H_1 d\bar{\theta} \tag{1.32}$$

and the oscillating part over $\bar{\theta}$ as

$$\{H_1\}_{\bar{\theta}} = H_1 - \langle \bar{H}_1 \rangle_{\bar{\theta}}. \tag{1.33}$$

Separating (1.31) into its average and oscillating parts yields, for $\bar{H}$ to first order,

$$\bar{H}(\bar{I}, \varepsilon\bar{\eta}, \varepsilon t) = H_0 + \varepsilon \langle \bar{H}_1 \rangle_{\bar{\theta}} \tag{1.34}$$

and, for S_1,

$$\omega \frac{\partial S_1}{\partial \bar{\theta}} = -\{H_1\}_{\bar{\theta}}, \tag{1.35}$$

which is easily integrated. To zero order, the adiabatic invariant is I. To first order, the new invariant is $\bar{I}$, which is given in terms of the old variables as

$$\bar{I}(I, \varepsilon\eta, \varepsilon t) = I - \varepsilon \frac{\partial S_1}{\partial \theta}. \tag{1.36}$$

Substituting (1.35) into (1.36) and writing θ for the dummy variable $\bar{\theta}$, one obtains

$$\bar{I} = I + \frac{\varepsilon\{H_1\}_\theta}{\omega}. \tag{1.37}$$

In fact, any function of $\bar{I}$ can be chosen as the adiabatic invariant.

1.1.2 Adiabatic Geometric Angle—Hannay Angle

In the phase space (p, q), the particle races around a track (i.e., a contour of the instantaneous Hamiltonian $H(p, q; t)$) of fixed area $2\pi I$ (with I being the action variable) but slowly changing shape. Given the rule of conservation of action I for the contour that the particle lies on, it seems natural to explore the development of the complementary variable, the angle variable, which describes the location of the particle on the contour; that is, one might ask how many circuits the particle has made [3].

When the Hamiltonian $H(p, q; t)$ is "frozen", the instantaneous frequency of motion of the particle that can be obtained can be expressed by the derivative $(2\pi)^{-1} dH/dI$. Thus, it is tempting to write the total angle traversed over time T as simply

$$\int_0^T \frac{dH(p(t), q(t); t)}{dI} dt = \int_0^T \frac{dH(I; t)}{dI} dt, \tag{1.38}$$

where, in the last form, H is considered a function $H(I; t)$ of the area of its contours and the adiabatic invariant and $I(t) = I$ (constant) is invoked. Since the angle variable can be changed by virtue of the changing (I, θ) coordinate system in phase space, this framework (1.38) is obviously incomplete. To reveal the true structure of the situation, it is necessary to interpret the time dependence of the Hamiltonian function (and the (I, θ) coordinate system) as being produced by carrying them along a path $\boldsymbol{R}(t)$ in a parameter space $\boldsymbol{R} \equiv (R_1, R_2, \ldots)$ of two or more dimensions in which the functions $H(p, q; \boldsymbol{R})$, $I(p, q; \boldsymbol{R})$, and $\theta(p, q; \boldsymbol{R})$ are uniquely defined. The point of making $\boldsymbol{R}$ more than one dimensional is that one wishes to consider closed evolutions $\boldsymbol{R}(T) = \boldsymbol{R}(0)$ in which $\boldsymbol{R}(t)$ forms a loop. With just one parameter (the length of the shortening pendulum, for instance), the only way to restore the original length is to reverse the shortening, in which case the holonomy effect is not realized.

The exact rates of change of a particle's action and angle in this framework are as follows:

$$\dot{I} = -\frac{\partial H}{\partial \theta} + \dot{\boldsymbol{R}} \cdot \frac{\partial I}{\partial \boldsymbol{R}} = \dot{\boldsymbol{R}} \cdot \frac{\partial I}{\partial \boldsymbol{R}}, \tag{1.39}$$

$$\dot{\theta} = \frac{\partial H}{\partial I} + \dot{\boldsymbol{R}} \cdot \frac{\partial \theta}{\partial \boldsymbol{R}}, \tag{1.40}$$

where the overdot denotes the time derivative. The last terms in both (1.39) and (1.40) are the rates of change of action and angle coordinates at a fixed point (p, q) in phase space. These equations for nonadiabatic evolution of the Hamiltonian lead to changes in both I and θ that depend on the trajectory selected, i.e., on the initial values of both I and θ. For adiabatic evolution, the equations become

$$\dot{I} = 0 + \dot{\boldsymbol{R}} \cdot \left\langle \frac{\partial I}{\partial \boldsymbol{R}} \right\rangle = 0, \tag{1.41}$$

$$\dot{\theta} = \frac{\partial H}{\partial I} + \dot{\boldsymbol{R}} \cdot \left\langle \frac{\partial \theta}{\partial \boldsymbol{R}} \right\rangle, \tag{1.42}$$

where the average brackets denote the average around the Hamiltonian contour on which the particle lies. For any function $f(p, q)$, one can define a function $\langle f \rangle$ of action I by

$$\langle f \rangle = \frac{1}{2\pi} \oint_{\text{contour through}(p,q)} f d\theta \equiv \frac{1}{2\pi} \int f(p, q) \delta(I(p, q) - I) dp dq. \tag{1.43}$$

The average in (1.41) vanishes identically by Liouville's theorem and yields $\dot{I} = 0$ as required. There is no reason, however, why the average in (1.42) should vanish; therefore, the integration of this equation gives the dynamical angle change anticipated in (1.38) plus the additional angular change (namely, $\Delta\theta$) that we are interested in:

$$\Delta\theta = \int \dot{\boldsymbol{R}} \cdot \left\langle \frac{\partial\theta(p(t), q(t); \boldsymbol{R}(t))}{\partial \boldsymbol{R}} \right\rangle dt = \int \left\langle \frac{\partial\theta}{\partial \boldsymbol{R}} \right\rangle \cdot d\boldsymbol{R}. \tag{1.44}$$

In the last expression, time t has been completely eliminated because by definition (1.43), the average is a function of a conserved parameter (i.e., the initial action I). A different field $\langle \partial\theta/\partial\boldsymbol{R}\rangle$ exists for each I, on which $\Delta\theta$ therefore depends. This parameter does not depend on the initial angle.

For fixed I, the field $\langle \partial\theta/\partial\boldsymbol{R}\rangle$ depends on the angle variable coordination $\theta(p, q; \boldsymbol{R})$, which is to some extent arbitrary. Unlike the lines of constant action $I(p, q; \boldsymbol{R})$, which, for fixed $\boldsymbol{R}$, are fully determined as the contours of the Hamiltonian $H(p, q; \boldsymbol{R})$, the lines of constant angle are specified only after one of them (say $\theta = 0$) is chosen. This one, and thus all the others, can be arbitrarily twisted into a spiral, for example. Thus, the angle variable change $\Delta\theta$ inevitably depends on the angle coordinates chosen for the initial and final parameters $\boldsymbol{R}(0)$ and $\boldsymbol{R}(T)$. Only if these coordinate systems are identical, which in turn requires $\boldsymbol{R}(0) = \boldsymbol{R}(T)$ (barring especially favorable circumstances), can one expect to make coordinate-independent statements about $\Delta\theta$. The evolutions must be closed loops.

1.1.3 Example I: One-Dimensional Harmonic Oscillator

As the first example, we introduce the one-dimensional harmonic oscillator. To show the general method, we calculate to first order the adiabatic invariant for the slowly varying linear oscillator [2], whose Hamiltonian is

$$H_{ho} = \frac{1}{2}g(\tau)p^2 + \frac{1}{2}f(\tau)q^2, \tag{1.45}$$

where the small parameter ε has been inserted using $\tau = \varepsilon t$ to order the perturbation series. To prepare the system, one can transform to the action-angle variables I and θ of $H_0 = H_{ho}(\varepsilon = 0)$. In treating the harmonic oscillator, we adopt the generating function $F(q, \theta, \tau)$ given by

$$F = \frac{1}{2}Rq^2 \cot\theta, \tag{1.46}$$

where $R(\tau) = \sqrt{f/g}$. Using $p = \partial F/\partial q$, $I = -\partial F/\partial\theta$, and $H(I, \theta, \tau) = H_{ho}(p, q, \tau) + \partial F(q, \theta, \tau)/\partial t$, one obtains $q = \sqrt{2I/R}\sin\theta$ and $p = \sqrt{2I/R}\cos\theta$. As a result, the transformed Hamiltonian is

$$H = \omega_0 I + \varepsilon \frac{1}{2}\frac{R'}{R} I \sin 2\theta, \tag{1.47}$$

where $\omega_0(\tau) = \sqrt{fg}$. The prime notation denotes differentiation with respect to τ. To zero order, the adiabatic invariant is just

$$I = \frac{H_0}{\omega_0} = \text{const..} \tag{1.48}$$

This result implies that the number of quanta $\hbar\omega_0$ is conserved as the frequency of oscillation slowly varies. To find the first-order invariant, we apply (1.37) to (1.47) and obtain

$$\bar{I} = I(1 + \varepsilon P \sin 2\theta) = \text{const.}, \tag{1.49}$$

with $P(\varepsilon t) = R'/(2\omega_0 R)$. This expression shows that to first order, I contains a small component oscillating at twice the frequency of the fast variable. One can verify the constancy of the quantity $\bar{I}$ by taking the time derivative of (1.49),

$$\dot{\bar{I}} = \dot{I} + \varepsilon \dot{P} I \sin 2\theta + 2\varepsilon P I \cos 2\theta + \mathcal{O}(\varepsilon^2), \tag{1.50}$$

where the overdot denotes d/dt. If one applies Hamilton's equations to (1.47), the first and third terms on the right cancel, leaving to first order in ε

$$\dot{\bar{I}} = \varepsilon \dot{P} I \sin 2\theta. \tag{1.51}$$

When the standard slow perturbation ordering $\dot{P} \sim \varepsilon P$ is used, $\dot{\bar{I}}$ is of order ε^2. Therefore, $\bar{I}$ is a first-order invariant.

1.1.4 Example II: Celestial Two-Body Problem

The two-body problem is a special case of the motion of a particle in a central force field. Because of the conservation of angular momentum, the motion occurs on an invariant plane [4]. In the plane polar coordinates, the Hamiltonian for the motion of the particle is

$$H = \frac{1}{2m}\left(p_r^2 + \frac{p_\theta^2}{r^2}\right) + U(r), \tag{1.52}$$

where m is the mass of the particle, the momentum components are $p_r = m\dot{r}$ and $p_\theta = mr^2\dot{\theta}$ (with the overdot denoting the time derivative), and $U(r)$ is the potential function for the central force field. The Hamilton-Jacobi equation reads

$$\frac{1}{2m}\left[\left(\frac{\partial S_r}{\partial r}\right)^2+\frac{1}{r^2}\left(\frac{\partial S_\theta}{\partial \theta}\right)^2\right]+U(r)=E, \tag{1.53}$$

where $S(r,\theta)=S_r(r)+S_\theta(\theta)$ is a variable-separated generating function. One can rewrite the above Hamilton-Jacobi equation as follows:

$$\left(\frac{\partial S_\theta}{\partial \theta}\right)^2=2mr^2\left[E-\frac{1}{2m}\left(\frac{\partial S_r}{\partial r}\right)^2-U(r)\right]. \tag{1.54}$$

The arbitrary choices of both θ and r require

$$\left(\frac{\partial S_r}{\partial r}\right)^2=2m\left[E-U(r)\right]-\frac{l^2}{r^2}, \tag{1.55}$$

$$\left(\frac{\partial S_\theta}{\partial \theta}\right)^2=l^2. \tag{1.56}$$

Applying the definitions of action-angle variables, i.e.,

$$I_r=\frac{1}{2\pi}\oint p_r dr=\oint\frac{\partial S_r}{\partial r}dr, \tag{1.57}$$

$$I_\theta=\frac{1}{2\pi}\oint p_\theta d\theta=\oint\frac{\partial S_\theta}{\partial \theta}d\theta, \tag{1.58}$$

one has

$$I_r=\frac{1}{2\pi}\int_0^{2\pi}\left[2m(E-U(r))-\frac{l^2}{r^2}\right]^{1/2}dr, \tag{1.59}$$

$$I_\theta=\frac{1}{2\pi}\int_0^{2\pi}ld\theta=l. \tag{1.60}$$

If the central force field takes the form $U(r)=-\mu/r$ with a constant quantity μ, then the action variable becomes

$$I_r=-l+\frac{\mu}{2}\sqrt{\frac{2m}{-E}}. \tag{1.61}$$

The corresponding Hamiltonian is given by

$$H=E=-\frac{m\mu^2}{2(I_r+I_\theta)^2}. \tag{1.62}$$

The frequency of particle motion in both the r and θ directions is

$$\omega = \frac{\partial H}{\partial I_r} = \frac{\partial H}{\partial I_\theta} = \frac{m\mu^2}{(I_r + I_\theta)^3}. \tag{1.63}$$

One finds that the motion of the particle in an inverse-square force field is simply due to the identical motion frequency in two directions. Thus, we can view the two-body motion with a Newton inverse-square gravity as a reduced motion. Substituting the familiar two-body elliptic motion energy $E' = E/m = -\mu/(2a)$ (with a being the semimajor axis) back into the equation for the action variable I_r, one has

$$I_r + I_\theta = \sqrt{\mu a} = L, \tag{1.64}$$

which implies conservation of the angular momentum L. Combining this variable with the frequency equation and Kepler's third law $n^2a^3 = \mu$, one has

$$\omega = n = \frac{2\pi}{T}, \tag{1.65}$$

where T is the period of the elliptic motion; thus, n is the angular speed. For typical celestial motions, T is often large, and thus, ω is very small. This fact implies that the motions are nearly adiabatic.

1.1.5 Example III: Foucault Pendulum

The Foucault pendulum provides a simple and effective example of the anholonomy present in an adiabatically cycled system because the parameter space used to describe its motion is the physical space in which it moves. The Foucault pendulum is commonly considered from a rectangular coordinate system (x, y, z) fixed to the rotating Earth with its origin at the pendulum bob in its rest position and its z-axis pointing outward from the Earth along the axis or rest orientation of the pendulum. The x- and y-axes point south and east, respectively. The pendulum is treated in the small-oscillation limit, and the fictitious centrifugal force proportional to the square of the angular frequency of the Earth (and hence very small) can be neglected. The Foucault pendulum is then characterized as a simple two-dimensional harmonic pendulum with an added Coriolis force. The Lagrangian for this system reads [5]

$$L = \frac{m}{2}(\dot{x}^2 + \dot{y}^2) - \frac{m\Omega^2}{2}(x^2 + y^2) + m\omega_z(x\dot{y} - y\dot{x}), \tag{1.66}$$

where the overdot denotes the time derivative, m is the mass of the pendulum bob, $\Omega = \sqrt{g/l}$ is the angular frequency of the unperturbed pendulum, with g being the acceleration of gravity and l being the length of the pendulum, and $\omega_z = \omega \cos\theta$ is the z-component of the angular frequency of the Earth (i.e., ω) at colatitude θ. Since the Earth's rotation is an adiabatic perturbation of the pendulum, the frequencies obey $\omega_z \ll \Omega$.

The symmetry about the pendulum axis allows a transformation to polar coordinates as follows: $x = \rho \cos \phi$ and $y = \rho \sin \phi$. The Lagrangian in polar coordinates becomes

$$L = \frac{m}{2}(\dot{\rho}^2 + \rho^2 \dot{\phi}^2) - \frac{m\Omega^2}{2}\rho^2 + m\omega_z \rho^2 \dot{\phi}. \tag{1.67}$$

The generalized momenta and the desired Hamiltonian in polar coordinates are

$$p_\rho = \frac{\partial L}{\partial \dot{\rho}} = m\dot{\rho}, \quad p_\phi = \frac{\partial L}{\partial \dot{\phi}} = m\rho^2(\dot{\phi} + \omega_z), \tag{1.68}$$

$$H = \frac{p_\rho^2}{2m} + \frac{p_\phi^2}{2m\rho^2} - \omega_z p_\phi + \frac{m}{2}(\Omega^2 + \omega_z^2)\rho^2. \tag{1.69}$$

Because ϕ is cyclic in the Hamiltonian and p_ϕ is a constant of the motion, the Hamilton-Jacobi equation is separable, and action-angle variable analysis is possible. In fact, one can write the Hamiltonian in terms of the action variables defined by

$$I_\phi = \frac{1}{2\pi} \oint p_\phi d\phi = p_\phi, \quad I_\rho = \frac{1}{2\pi} \oint p_\rho d\rho. \tag{1.70}$$

The desired form of the Hamiltonian in terms of the action variables determined by the energy E is

$$H = E = (2I_\rho + |I_\phi|)\sqrt{\Omega^2 + \omega_z^2} - I_\phi \omega_z. \tag{1.71}$$

Since the angle variables are always cyclic in an action-angle Hamiltonian, the Hamilton equations for the action variables show these variables to be constants of the motion. The fundamental frequencies of the system are determined by the Hamilton equations for the angle variables.

For clockwise motion, the angular momentum of the bob is negative; thus, one has $\partial |I_\phi|/\partial I_\phi = -1$. The Hamilton equation for ϕ is

$$\dot{\phi} = \frac{\partial H}{\partial I_\phi} = -\sqrt{\Omega^2 + \omega_z^2} - \omega_z = -\omega_1, \tag{1.72}$$

yielding upon integration

$$\phi = -\omega_1(t - t_1). \tag{1.73}$$

For counter-clockwise motion, the angular momentum of the bob is positive; thus, $\partial |I_\phi|/\partial I_\phi = 1$. The Hamilton equation for ϕ is

$$\dot{\phi} = \frac{\partial H}{\partial I_\phi} = \sqrt{\Omega^2 + \omega_z^2} - \omega_z = \omega_2, \tag{1.74}$$

yielding upon integration

$$\phi = \omega_2(t - t_2). \tag{1.75}$$

Once the fundamental frequencies of the system have been calculated, one can examine the effect of a cyclic adiabatic change on the angular precession of the bob by computing the Hannay angle. The Hannay angle for both normal modes can be found by choosing the integration constants t_1 and t_2 such that $\phi = 2\pi$ at $t = 0$ and by then substituting the time for one revolution of the Earth (the cycle of the adiabatic perturbation) $t = 2\pi/\omega$ to obtain

$$\phi = \pm\frac{2\pi}{\omega}\sqrt{\Omega^2 + \omega^2\cos^2\theta} + 2\pi(1 - \cos\theta). \tag{1.76}$$

The first term is the dynamic phase of either of the modes, which is interpretable as the angle swept out in the adiabatic limit by a simple conical pendulum in one day if it were fixed to a nonrotating frame. The second term is the geometric phase (or the Hannay angle), which is dependent only on the colatitude of the bob and is independent of both the initial conditions imposed on the bob and the duration of the adiabatic perturbation. The Hannay angle is the solid angle of the cone of the half angle θ swept out by the pendulum axis during one day.

1.2 Quantum Adiabatic Evolution

1.2.1 Quantum Adiabatic Theorem

A neutron or an atom moving in an inhomogeneous field is acted on by a time-varying field in the reference system of the particle. If the variation in the field is sufficiently slow, according to the adiabatic theorem, then the particle remains in the same state with respect to the instantaneous value of the field [6]. The adiabatic theorem is an important concept in quantum mechanics. Its original form, as formulated by Born and Fock, was stated as follows [7]:

"A physical system remains in its instantaneous eigenstate if a given perturbation is acting on it slowly enough and if there is a gap between the eigenvalue and the rest of the Hamiltonian's spectrum."

In simpler terms, a quantum mechanical system subjected to gradually changing external conditions adapts its functional form. However, when subjected to rapidly varying conditions, there is insufficient time for the functional form to adapt, so the spatial probability density remains unchanged.

The adiabatic theorem, as a fundamental theorem in quantum mechanics, plays a crucial role in our understanding and manipulation of the microscopic world

[8–12]. Recent years have witnessed its growing importance in the quantum control of newly formed matter—the Bose-Einstein condensate [13] and adiabatic quantum computation [14]. However, Marzlin and Sanders warned that application of the theorem can lead to inconsistency [15]. A subsequent work [16] explicitly formulated a "counter-example" with a two-level model illustrating that the adiabatic condition that is widely recognized and commonly used is not sufficient to ensure the accuracy of the adiabatic approximation.

In the present section, we formulate the quantum adiabatic evolution within a parameter domain rather than the time domain [17]. Within this new formulation, we can state the quantum adiabatic theorem consistently and investigate the fidelity of the adiabatic approximation quantitatively, naturally avoiding the above issues.

The system we consider is a Hamiltonian containing slowly varying dimensionless parameters $\boldsymbol{R}(t)$ belonging to a given regime $[\boldsymbol{R}_0, \boldsymbol{R}_1]$, say, $H(\boldsymbol{R}(t))$. We initially have a state, for example, the ground state $|E_0(\boldsymbol{R}(t_0))\rangle$ with energy $E_0(\boldsymbol{R}(t_0))$. The wave function $|\Psi(t)\rangle$ fulfills the usual Schrödinger equation, i.e., $i(d|\Psi\rangle/dt) = H(\boldsymbol{R}(t))|\Psi(t)\rangle$ with $\hbar = 1$. The above problem has a well-known adiabatic approximate solution:

$$|\Psi_{ad}\rangle = \mathrm{e}^{-i\int^t E_0 dt}\mathrm{e}^{i\lambda_0}|E_0(\boldsymbol{R}(t))\rangle, \tag{1.77}$$

where $\lambda_0 = i\int^t dt\langle E_0|\dot{E}_0\rangle$ is the geometric phase term [18] with the overdot denoting the time derivative. The above equation is the explicit formulation of the adiabatic theorem stating that the initial nondegenerate ground state remains the instantaneous ground state and evolves only in its phase, given by the time integral of the eigenenergy (known as the dynamical phase) and a quantity independent of the time duration (known as the geometric phase).

The problems is to determine how close the above adiabatic approximate solution is to the actual solution $|\Psi(t)\rangle$. To clarify this question and formulate it quantitatively, we introduce two physical quantities, namely, the adiabatic parameter and the adiabatic fidelity.

The dimensionless adiabatic parameter is defined as the ratio between the change rate of the external parameters and the internal characteristic time scale of the quantum system (i.e., the Rabi frequency $|E_m - E_n|$), used to measure how slowly the external parameter changes with time:

$$\varepsilon = \max\frac{|\dot{\boldsymbol{R}}|}{|E_n(\boldsymbol{R}) - E_m(\boldsymbol{R})|}, \quad m \neq n. \tag{1.78}$$

The case of $\varepsilon \to 0$ corresponds to the adiabatic limit. Following this definition, the adiabatic condition can be written as ($\hbar = 1$ throughout)

$$\max|\dot{\boldsymbol{R}}| \ll |E_n(\boldsymbol{R}) - E_m(\boldsymbol{R})|, \quad m \neq n. \tag{1.79}$$

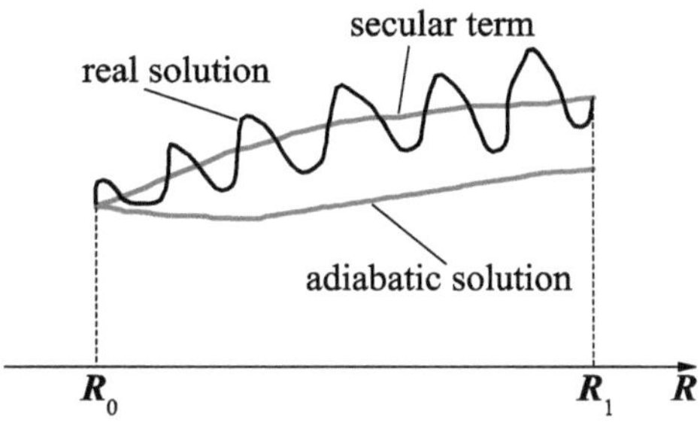

Fig. 1.1 Schematic illustration of the quantum adiabatic evolution formulated in the parameter domain; see the text for a detailed description

The adiabatic fidelity is introduced to measure how close the adiabatic solution is to the actual one, $F_{ad} = |\langle\Psi(t)|\Psi_{ad}\rangle|^2$. The convergence of the adiabatic fidelity to unity uniformly over the range $\boldsymbol{R} \in [\boldsymbol{R}_0, \boldsymbol{R}_1]$ in the adiabatic limit ($\varepsilon \to 0$) indicates the validity of the adiabatic approximation. Evaluation of the fidelity function provides an estimate of how satisfactory the adiabatic approximation is.

In Fig. 1.1, we schematically illustrate the physical process described above. Our main result is that the distance between the adiabatic solution and the actual one comprises two parts: a fast oscillation term and a secular term. The time scale of the oscillation is the Rabi period, and its amplitude is proportional to the square of the adiabatic parameter. The amplitude of the secular term is exponentially small ($\sim e^{-1/\varepsilon}$), supposing that the Berry connections of the system are regular, and transitions to a power-law ($\sim\varepsilon^x$, $x < 2$) if the Berry connections have singularity or if the external parameters vary in time nonlinearly.

We start our statement by writing the wavefunction as a superposition of the instantaneous eigenstates

$$|\Psi(t)\rangle = \sum_n C_n(t) e^{-i\int^t dt(E_n - i\langle E_n(\boldsymbol{R})|\dot{E}_n(\boldsymbol{R})\rangle)} |E_n(\boldsymbol{R}(t))\rangle, \tag{1.80}$$

with the overdot denoting the time derivative, and we suppose that the initial state is the ground state, i.e., $C_0(t=0) = 1$, $C_n(t=0) = 0$, with $n \neq 0$. Then, the adiabatic approximate solution takes the form of (1.77), and the adiabatic fidelity is $F_{ad} = |\langle\Psi(t)|\Psi_{ad}\rangle|^2 = |C_0|^2 \sim 1 - |\Delta C_n|^2$, where $n \neq 0$. To evaluate the adiabatic fidelity, we need to quantitatively evaluate the change of the coefficients C_n with respect to time.

Substituting the above solution into the Schrödinger equation, we have the following differential equation for the coefficients:

$$\frac{d}{dt}C_n = i \sum_{m\neq n} e^{i\int^t((E_n - \alpha_{nn}\dot{\boldsymbol{R}}) - (E_m - \alpha_{mm}\dot{\boldsymbol{R}}))dt} \alpha_{nm}(\boldsymbol{R}) \frac{d\boldsymbol{R}}{dt} C_m, \tag{1.81}$$

where $\alpha_{nm}(\boldsymbol{R})$ are the Berry connections. Both the off-diagonal and diagonal Berry connections have clear physical meaning and important applications [19]. We first suppose that these Berry connections and the gradient of the instantaneous

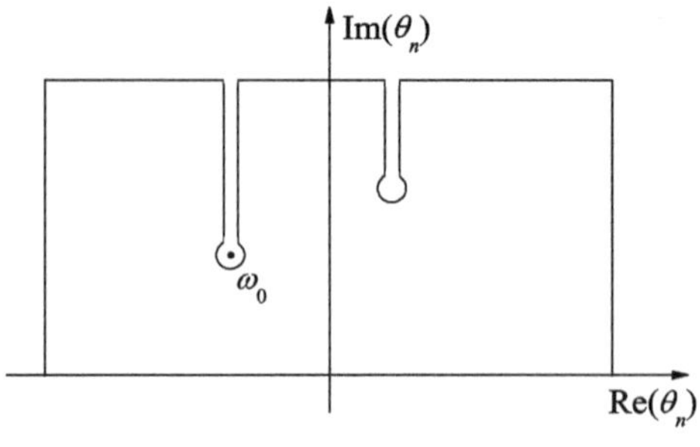

Fig. 1.2 Integral paths and singular points in the complex plane

eigenenergies are not singular (NS) as functions of the external parameters, i.e.,

$$\alpha_{nm}(\boldsymbol{R}) = \langle E_n | i\nabla_{\boldsymbol{R}} | E_m \rangle; \quad \beta_n(\boldsymbol{R}) = \nabla_{\boldsymbol{R}} E_n(\boldsymbol{R}), \quad \text{NS}. \tag{1.82}$$

The right-hand side of (1.81) contains the unknown C_m. To the first order approximation, we take $C_0 = 1$ and $C_m = 0$, $m \neq 0$ on the right-hand side of (1.81). Then, Eq. (1.81) shows that the change consists of two parts: the fast oscillation term and the secular term. The time scale of the oscillation is the Rabi period, whose amplitude is proportional to the adiabatic parameter under the condition that the Berry connections are regular with limitation. In contrast, the secular term may be exponentially small and of the form $e^{-1/\varepsilon}$ or of a power-law form depending on the Berry connections, as shown below.

Let us denote $\theta_n = \int^t (E_n - E_0)dt$; the upper bound of the increment on the coefficients ($n \neq 0$) can then be evaluated as follows:

$$\Delta C_n \sim \int_{\theta_n(\boldsymbol{R}_0)}^{\theta_n(\boldsymbol{R}_1)} \frac{e^{i\theta_n}}{E_n - E_0} \alpha_{n0} \dot{\boldsymbol{R}} d\theta_n \tag{1.83}$$

$$= \int_{-\infty}^{\infty} \cdots d\theta_n - \left(\int_{-\infty}^{\theta_n(\boldsymbol{R}_0)} + \int_{\theta_n(\boldsymbol{R}_1)}^{\infty} \right) \cdots d\theta_n, \tag{1.84}$$

where we set on the right-hand side of (1.81) the coefficients $C_m \sim 0$ for $m \neq 0$ and $C_0 \sim 1$ since we wish to estimate the upper bound of the adiabatic approximation. For simplicity and without loss of generality, in the following deductions, we regard the slow-varying parameter as a scalar quantity, and we assume that $dR/dt \sim \varepsilon g(R)$, with the function $g(R)$ being regular with limitation.

The first term on the right-hand side of (1.84) is an infinite integral that can be estimated by extending the integral to the upper half-plane with a closed path at infinity, as shown in Fig. 1.2. The integral along the upper horizontal path of the closure is zero because $e^{-\mathrm{Im}\theta_n} \to 0$ there. Furthermore, the integrals along the vertical paths also vanish because of the fast oscillation of the function $e^{-i\mathrm{Re}\theta_n} \to 0$ at infinity [1]. Hence, the main contribution to the first term comes from the pole point, $\theta_n^c = \int^{t_c} (E_n - E_0)dt \sim \varepsilon^{-1} \int^{R_c} [(E_n - E_0)/g(R)]dR$, determined by the equation $E_n(\theta_n^c) - E_0(\theta_n^c) = 0$. Under the assumption of nondegeneracy, the solutions of the above equation are complex with nonzero imaginary parts. Let ω_0 be the singularity closest to the real axis, i.e., the one with the smallest (positive) imaginary part (see Fig. 1.2). Then, the first term is approximately bounded by $e^{-\mathrm{Im}\omega_0} \sim e^{-\varepsilon^{-1}|\int^{|\mathrm{Im}(R_c)|}[(E_n-E_0)/g(R)]dR|}$, which contributes to the secular term with an exponentially small quantity in the adiabatic limit.

The second term on the right-hand side of (1.84) depends on the boundary condition. If the boundary values are large enough, the terms in the parentheses of (1.84) will be small relative to the first term. For an infinite boundary of the parameter domain, such as that in the well-known Landau-Zener model [20], i.e., $R_{1,0} \to \pm\infty$, the integral vanishes because $\theta_n(R_{1,0}) = \pm\infty$; for a finite boundary condition, we obtain a quantity of order ε.

Now, we consider the situation that the Berry connection contains a singularity of form $1/R^\alpha$ at $R(t^*) = 0$. We then evaluate the above integral in the neighborhood domain $[-\Delta t + t^*, \Delta t + t^*]$ of the singular point; the integral over other regimes is regular and contributes a quantity of order ε. Near the singular point,

$$
\begin{aligned}
|\Delta C_n| &\sim \left| \int_{-\Delta t+t^*}^{\Delta t+t^*} e^{i\int^t (E_n-E_0)dt} \alpha_{n0}(R)dR \right| \\
&= \left| \int_{\Delta R_-}^{\Delta R_+} e^{i\int^t (E_n-E_0)dt} \alpha_{n0}(R)dR \right| \\
&\sim \varepsilon^{(1-\sigma)}.
\end{aligned} \tag{1.85}
$$

In the above deduction, we have taken advantage of the relation $\Delta R_\pm = R(\pm\Delta t + t^*) \propto \pm\varepsilon$.

The situation is divided into two cases: $\sigma < 0$ and $\sigma \geq 0$. For $\sigma < 0$, this type of singularity can be removed because the integral is finite. The integral in the neighborhood domain $[-\varepsilon, \varepsilon]$ of the singularity contributes a quantity of order $\varepsilon^{(1-\sigma)}$. We thus expect that the adiabatic fidelity approaches unity uniformly in the $2(1-\sigma)$ power law of the adiabatic parameter, i.e.,

$$
1 - F_{ad} \sim \varepsilon^{2(1-\sigma)}. \tag{1.86}
$$

For the case of $\sigma \geq 0$, the singularity is irremovable, and the adiabatic approximation is expected to break down.

The above discussion is readily extended to the case that the slow-varying parameters change nonlinearly with time, i.e., $R = \varepsilon t^\sigma$, where σ is any positive number. The nonlinear time-dependent parameter has multiple physical origins; in the molecule spin system, for example, the effective field varies nonlinearly in time [21]. Another field with a broad number of examples is quantum optics, in which the Rabi frequency coupling different levels (i.e., stimulated Raman adiabatic passage) often depends nonlinearly on time [22]. Here, we suppose that the Berry connections of a quantum system are regular with limitation as a function of the parameter R and that the level spacings are of order 1. To apply our theory, we introduce the new parameters

R' and ε' through the expressions $\varepsilon' = \varepsilon^{1/\sigma}$ and $R' = \varepsilon' t$. Consequently, $R = R'^{\sigma}$. As a function of the new parameter R', the singularity of the Berry connections is determined by $dR/dR' \sim 1/R'^{1-\sigma}$. Our discussion is divided into two cases: (i) $\sigma > 1$ and (ii) $\sigma < 1$. In the former case, the Berry connections as functions of the new parameter are regular, so the adiabatic fidelity is determined by the short-term oscillation and is expected to converge to one in a power law of ε'^2. Then, we have

$$1 - F_{ad} \sim \varepsilon^{\frac{2}{\sigma}}. \tag{1.87}$$

In the latter case, the Berry connections as functions of the new parameter are singular, of the type $1/R'^{1-\sigma}$. Fortunately, this singularity is removable, resulting in an upper bound of the adiabatic fidelity of $\varepsilon'^{2\sigma}$, i.e.,

$$1 - F_{ad} \sim \varepsilon^2. \tag{1.88}$$

Note that in this case, the upper bound of the adiabatic fidelity is independent of the nonlinear index σ.

1.2.2 Adiabatic Geometric Phase—Berry Phase

A quantum system in an eigenstate, slowly transported around a circuit by varying parameters in its Hamiltonian, in addition to the familiar dynamical phase factor, ultimately acquires a geometric phase factor [18].

Let the Hamiltonian H be changed by varying parameters $\boldsymbol{R} = (R_1, R_2, \ldots)$, on which it depends. Then, the evolution of the system between times $t = 0$ and $t = T$ can be interpreted as movement around a closed path $\boldsymbol{R}(t)$ in parameter space with the Hamiltonian $H(\boldsymbol{R}(t))$ and such that $\boldsymbol{R}(T) = \boldsymbol{R}(0)$. This path is henceforth called a circuit and denoted by C. For the adiabatic approximation to apply, T must be large enough.

The quantum state $|\psi(t)\rangle$ of the system evolves according to Schrödinger's equation,

$$H(\boldsymbol{R}(t))|\psi(t)\rangle = i\hbar|\dot{\psi}(t)\rangle, \tag{1.89}$$

where the overdot denotes the time derivative. At any instant, the natural basis consists of the eigenstates $|n(\boldsymbol{R})\rangle$ (assumed to be discrete) of $H(\boldsymbol{R})$ for $\boldsymbol{R} = \boldsymbol{R}(t)$ that satisfy

$$H(\boldsymbol{R})|n(\boldsymbol{R})\rangle = E_n(\boldsymbol{R})|n(\boldsymbol{R})\rangle, \tag{1.90}$$

with energies $E_n(\boldsymbol{R})$. This eigenvalue equation implies that no relation exists between the phases of the eigenstates $|n(\boldsymbol{R})\rangle$ at different $\boldsymbol{R}$. For present purposes, any (differentiable) choice of phases can be made, provided that $|n(\boldsymbol{R})\rangle$ is single-valued in a parameter domain that includes the circuit C.

Adiabatically, a system prepared in one of these states $|n(\boldsymbol{R}(0))\rangle$ evolves with H and is thus in the state $|n(\boldsymbol{R}(t))\rangle$ at t. Therefore, $|\psi\rangle$ can be written as

$$|\psi(t)\rangle = \mathrm{e}^{-\frac{i}{\hbar}\int_0^t dt' E_n(\boldsymbol{R}(t'))} \mathrm{e}^{i\lambda_n(t)} |n(\boldsymbol{R}(t))\rangle. \tag{1.91}$$

The first exponential is the familiar dynamical phase factor. The object of attention is the second exponential. The crucial point is that its phase $\lambda_n(t)$ is nonintegrable; λ_n cannot be written as a function of $\boldsymbol{R}$ and in particular is not single-valued under continuation around a circuit, i.e., $\lambda_n(T) \neq \lambda_n(0)$.

The function $\lambda_n(t)$ is determined by the requirement that $|\psi(t)\rangle$ satisfies Schrödinger's equation; direct substitution of (1.91) into (1.89) leads to

$$\dot{\lambda}_n(t) = i\langle n(\boldsymbol{R}(t))|\nabla_{\boldsymbol{R}} n(\boldsymbol{R}(t))\rangle \cdot \dot{\boldsymbol{R}}(t). \tag{1.92}$$

The total phase change of the state $|\psi\rangle$ around C is

$$|\psi(T)\rangle = \mathrm{e}^{-\frac{i}{\hbar}\int_0^T dt E_n(\boldsymbol{R}(t))} \mathrm{e}^{i\lambda_n(C)} |\psi(0)\rangle, \tag{1.93}$$

where the geometric phase change is given by

$$\lambda_n(C) = i \oint_C \langle n(\boldsymbol{R})|\nabla_{\boldsymbol{R}} n(\boldsymbol{R})\rangle \cdot d\boldsymbol{R}. \tag{1.94}$$

Thus, $\lambda_n(C)$ is expressed as a circuit integral in parameter space and is independent of how the circuit is traversed (provided of course that this movement is slow enough for the adiabatic approximation to hold). The normalization of the state $|n\rangle$ implies that $\langle n|\nabla_{\boldsymbol{R}} n\rangle$ is imaginary, which guarantees that λ_n is real.

Direct evaluation of $|\nabla_{\boldsymbol{R}} n\rangle$ requires a locally single-valued basis for $|n\rangle$. This difficulty can be avoided by transforming the circuit integral (1.94) into a surface integral over any surface in parameter space whose boundary is C. To apply familiar vector calculus, the parameter space is taken as three-dimensional, and this condition turns out to be the most important case in applications.

Applying Stokes's theorem to (1.94), one has

$$\lambda_n(C) = -\mathrm{Im} \int\!\!\int_C d\boldsymbol{S} \cdot \sum_{m \neq n} \langle \nabla n|m\rangle \times \langle m|\nabla n\rangle, \tag{1.95}$$

where $d\boldsymbol{S}$ is the area element in $\boldsymbol{R}$ space and the exclusion in the summation is justified by $\langle n|\nabla n\rangle$ being imaginary. The off-diagonal elements are obtained from (1.90) as follows:

$$\langle m|\nabla n\rangle = \frac{\langle m|\nabla H|n\rangle}{E_n - E_m}, \quad m \neq n. \tag{1.96}$$

Therefore, λ_n can be rewritten as

$$\lambda_n(C) = -\int\int_C d\boldsymbol{S} \cdot \boldsymbol{B}_n(\boldsymbol{R}), \tag{1.97}$$

where

$$\boldsymbol{B}_n(\boldsymbol{R}) \equiv \mathrm{Im} \sum_{m \neq n} \frac{\langle n(\boldsymbol{R})|\nabla_{\boldsymbol{R}} H(\boldsymbol{R})|m(\boldsymbol{R})\rangle \times \langle m(\boldsymbol{R})|\nabla_{\boldsymbol{R}} H(\boldsymbol{R})|n(\boldsymbol{R})\rangle}{(E_m(\boldsymbol{R}) - E_n(\boldsymbol{R}))^2}. \tag{1.98}$$

Clearly, $\lambda_n(C)$ is zero for a circuit that retraces itself and so encloses no area.

Equations (1.97) and (1.98) are the central results. Since the dependence on $|\nabla n\rangle$ has been eliminated, the phase relations between eigenstates with different parameters are immaterial, and (as is evident from the form of (1.98)) it is no longer necessary to choose $|m\rangle$ and $|n\rangle$ to be single-valued in $\boldsymbol{R}$: any solutions of (1.90) may be employed without affecting the value of $\boldsymbol{B}_n$. This is a surprising conclusion, as can be seen by comparing (1.97) with (1.95), which shows that $\boldsymbol{B}_n$ is the curl of a vector, $\langle n|\nabla n\rangle$, and that this vector certainly does depend on the choice of the phase of the (single-valued) eigenstate $|n(\boldsymbol{R})\rangle$. The dependence on the phase is of the following kind: if $|n\rangle \to e^{i\mu(\boldsymbol{R})}|n\rangle$, then $\langle n|\nabla n\rangle \to \langle n|\nabla n\rangle + i\nabla\mu$ (in another context, the importance of such gauge transformations was emphasized by Wu and Yang [23]). Hence, the vector is not unique, but its curl is. The quantity $\boldsymbol{B}_n$ is analogous to a "magnetic field" (in parameter space) whose "vector potential" is $\mathrm{Im}\langle n|\nabla n\rangle$. As described elsewhere (1.98), $\nabla \cdot \boldsymbol{B}_n$ vanishes, thus confirming that (1.97) gives a unique value for $\lambda_n(C)$.

1.2.3 Virtual Magnetic Monopole

The denominators in (1.98) imply that if the circuit C lies close to a point $\boldsymbol{R}^*$ in parameter space at which the state $|n\rangle$ is involved in a degeneracy, then $\boldsymbol{B}_n(\boldsymbol{R})$ and, thus, $\lambda_n(C)$ are dominated by the terms m corresponding to the other states involved. We discuss the most common situation, where the degeneracy involves only two states [18], to be denoted by $|\pm(\boldsymbol{R})\rangle$, where $E_+(\boldsymbol{R}) \geq E_-(\boldsymbol{R})$. For $\boldsymbol{R}$ near $\boldsymbol{R}^*$, $H(\boldsymbol{R})$ can be expanded to first order in $(\boldsymbol{R} - \boldsymbol{R}^*)$, and one has

$$\boldsymbol{B}_+(\boldsymbol{R}) = \mathrm{Im} \frac{\langle +(\boldsymbol{R})|\nabla_{\boldsymbol{R}} H(\boldsymbol{R})| - (\boldsymbol{R})\rangle \times \langle -(\boldsymbol{R})|\nabla_{\boldsymbol{R}} H(\boldsymbol{R})| + (\boldsymbol{R})\rangle}{(E_+(\boldsymbol{R}) - E_-(\boldsymbol{R}))^2}. \tag{1.99}$$

Clearly, one also has $\boldsymbol{B}_-(\boldsymbol{R}) = -\boldsymbol{B}_+(\boldsymbol{R})$, and thus, $\lambda_-(C) = -\lambda_+(C)$.

Without loss of generality, one can set $E_\pm(\boldsymbol{R}^*) = 0$ and $\boldsymbol{R}^* = 0$. Indeed, $H(\boldsymbol{R})$ can be represented as a 2×2 Hermitian matrix coupling the two states. The most general such matrix satisfying the given conditions depends on three parameters: x, y, and z, which are taken as the components of $\boldsymbol{R}$ and by linear transformation in $\boldsymbol{R}$

space can be brought into the following form:

$$H(\boldsymbol{R}) = \frac{1}{2}\begin{pmatrix} z & x - iy \\ x + iy & -z \end{pmatrix}. \tag{1.100}$$

The eigenvalues are given by

$$E_{\pm}(\boldsymbol{R}) = \pm\frac{1}{2}(x^2 + y^2 + z^2)^{1/2} = \pm\frac{1}{2}R. \tag{1.101}$$

Therefore, the degeneracy is an isolated point at which all three parameters vanish. This result corresponds to a well-known result of Von Neumann and Wigner [24]: for general Hamiltonians (Hermitian matrices), it is necessary to change three parameters to cause a degeneracy to occur accidentally, that is, not on account of symmetry. Alternatively stated, degeneracies have co-dimension three.

The form (1.100) was chosen to exploit the fact that

$$\nabla H = \frac{1}{2}\boldsymbol{\sigma}, \tag{1.102}$$

where σ_x, σ_y, and σ_z denote the Pauli spin matrices, which are the components of the vector operator $\boldsymbol{\sigma}$. When computing the matrix elements in (1.99), one can greatly simplify the calculations by taking advantage of the isotropy of spin and temporarily rotating the axes (so that the z-axis points along $\boldsymbol{R}$) and by employing the following relations, which come from the commutation laws between the components of $\boldsymbol{\sigma}$:

$$\sigma_x|\pm\rangle = |\mp\rangle, \quad \sigma_y|\pm\rangle = \pm i|\mp\rangle, \quad \sigma_z|\pm\rangle = \pm|\pm\rangle. \tag{1.103}$$

With these rotated axes, Eq. (1.99) gives

$$\begin{cases} B_{x+} = \frac{\mathrm{Im}\langle+|\sigma_y|-\rangle\langle-|\sigma_z|+\rangle}{2R^2} = 0, \\ B_{y+} = \frac{\mathrm{Im}\langle+|\sigma_z|-\rangle\langle-|\sigma_x|+\rangle}{2R^2} = 0, \\ B_{z+} = \frac{\mathrm{Im}\langle+|\sigma_x|-\rangle\langle-|\sigma_y|+\rangle}{2R^2} = \frac{1}{2R^2}. \end{cases} \tag{1.104}$$

Reverting to unrotated axes, one has

$$\boldsymbol{B}_{+}(\boldsymbol{R}) = \frac{\boldsymbol{R}}{2R^3}. \tag{1.105}$$

The use of (1.97) illustrates that the phase change $\lambda_{+}(C)$ is the flux through C of the magnetic field of a monopole with strength $-\frac{1}{2}$ located at the degeneracy. Hence, for the natural choice (1.100) of the standard form for H, one can obtain the satisfying result that the geometric phase factor associated with C is given by

$$\mathrm{e}^{i\lambda_{\pm}(C)} = \mathrm{e}^{\mp i\Omega(C)/2}, \tag{1.106}$$

where $\Omega(C)$ is the solid angle that C subtends at the degeneracy, that is, a measure of the view of the circuit as seen from the degeneracy. Since Ω can change only in multiples of 4π (when the surface is deformed to pass through the degeneracy), the phase factor is independent of the choice of surface spanning C.

1.2.4 Nonadiabatic Geometric Phase—Aharonov-Anandan Phase

In the preceding sections, we introduced the Berry phase [18] in the quantum adiabatic process and its classical analog—the Hannay angle [3]. By relaxing the adiabatic approximation, these concepts were then generalized by Aharonov and Anandan [25] to study the phase associated with a cyclic evolution (which occurs when a state returns to its initial condition) in quantum mechanics.

The significance of Aharonov and Anandan's generalization is twofold. On the one hand, the cyclic evolution of a physical system, both experimentally and theoretically, is of high interest in physics. On the other hand, the universal existence of the cyclic evolution is guaranteed for any quantum system. This condition can be easily recognized by considering the eigenvectors of the unitary evolution operator for a quantum system. An explicit example is the time-periodic Hamiltonian system where the Floquet theorem applies. The eigenfunctions of the Floquet operator, which are referred to as Bloch wave functions in condensed matter physics, are cyclic solutions of great interest in physics.

The Aharonov-Anandan (AA) phase [25] is a universal geometric phase in the sense that it is the same for an infinite number of possible motions along the curves in the Hilbert space $\mathscr{H}$ that project to a given closed curve $\hat{C}$ in the projective Hilbert space $\mathscr{P}$ of rays of $\mathscr{H}$ and the possible Hamiltonians $H(t)$ that propagate the state along these curves. We assume that the normalized state $|\psi(t)\rangle \in \mathscr{H}$ evolves according to the Schrödinger equation,

$$i\hbar\frac{d}{dt}|\psi(t)\rangle = H(t)|\psi(t)\rangle, \tag{1.107}$$

such that $|\psi(\tau)\rangle = \mathrm{e}^{i\phi}|\psi(0)\rangle$, where ϕ is real. Let Π, i.e., $\mathscr{H} \to \mathscr{P}$, be the projection map defined by $\Pi(|\psi\rangle) = \{|\psi'\rangle : |\psi'\rangle = c|\psi\rangle, c \text{ is a complex number}\}$. Then, $|\psi(t)\rangle$ defines a curve C: $[0, \tau] \to \mathscr{H}$, with $\hat{C} \equiv \Pi(C)$ being a closed curve in $\mathscr{P}$. Conversely, given any such curve C, one can define a Hamiltonian function $H(t)$ so that (1.107) is satisfied for the corresponding normalized $|\psi(t)\rangle$. Now define $|\tilde{\psi}(t)\rangle = \mathrm{e}^{-i\lambda(t)}|\psi(t)\rangle$ such that $\lambda(\tau) - \lambda(0) = \phi$. Then, $|\tilde{\psi}(\tau)\rangle = |\tilde{\psi}(0)\rangle$, and from (1.107), one has

$$-\frac{d\lambda}{dt} = \frac{1}{\hbar}\langle\psi(t)|H|\psi(t)\rangle - \langle\tilde{\psi}(t)|i\frac{d}{dt}|\tilde{\psi}(t)\rangle. \tag{1.108}$$

Thus, if one removes the dynamical part from the phase ϕ by defining

$$\alpha \equiv \phi + \frac{1}{\hbar}\int_0^{\tau} \langle\psi(t)|H|\psi(t)\rangle dt, \tag{1.109}$$

then it follows from (1.108) that

$$\alpha = \int_0^{\tau} \langle\tilde{\psi}(t)|i\frac{d}{dt}|\tilde{\psi}(t)\rangle dt. \tag{1.110}$$

It is clear that the same $|\tilde{\psi}(t)\rangle$ can be chosen for every curve C for which $\Pi(C) = \hat{C}$ through an appropriate choice of $\lambda(t)$. Therefore, α, defined by (1.109), called the AA phase, is independent of ϕ and H for a given closed curve $\hat{C}$. Indeed, for a given $\hat{C}$, $H(t)$ can be chosen so that the second term in (1.109) is zero, which may be regarded as an alternative definition of α. In addition, from (1.110), α is independent of the parameter t of $\hat{C}$ and is uniquely defined up to $2n\pi$ ($n =$integer). Thus, $e^{i\alpha}$ is a geometric property of the unparameterized image of $\hat{C}$ in $\mathscr{P}$ only.

Consider a slowly varying $H(t)$, with $H(t)|n(t)\rangle = E_n(t)|n(t)\rangle$, for a complete set $\{|n(t)\rangle\}$. If one writes

$$|\psi(t)\rangle = \sum_n c_n(t)\exp\left(-\frac{i}{\hbar}\int E_n dt\right)|n(t)\rangle \tag{1.111}$$

and uses (1.107) and the time derivative of the eigenvector equation, one has

$$\dot{c_m} = -c_m\langle m|\dot{m}\rangle - \sum_{n\neq m} c_n \frac{\langle m|\dot{H}|n\rangle}{E_n - E_m}\exp\left[\frac{i}{\hbar}\int (E_m - E_n)dt\right], \tag{1.112}$$

where the overdot denotes the time derivative. Assume that

$$\sum_{n\neq m}\left|\frac{\hbar\langle m|\dot{H}|n\rangle}{(E_n - E_m)^2}\right| \ll 1. \tag{1.113}$$

Then, if $c_n(0) = \delta_{nm}$, the last term in (1.112) is negligible, and the system therefore continues as an eigenstate of $H(t)$ to satisfactory approximation. In this adiabatic approximation, (1.112) yields

$$c_m(t) \simeq \exp\left(-\int\langle m|\dot{m}\rangle dt\right)c_m(0). \tag{1.114}$$

For a cyclic adiabatic evolution, the phase $i\int_0^{\tau}\langle m|\dot{m}\rangle dt$ is independent of the chosen $|m(t)\rangle$; Berry regarded this property as a geometrical property of the parameter space of which H is a function. However, this phase is the same as (1.110) on choosing $|\tilde{\psi}(t)\rangle \simeq |m(t)\rangle$ in the present approximation. The parameter α, defined by (1.109), does not depend on any approximation, and thus, (1.110) is exactly valid. Moreover, $|\psi(t)\rangle$ need not be an eigenstate of $H(t)$, unlike in the limiting case studied by Berry. In addition, it is neither necessary nor sufficient to go around a (nontrivial) closed curve in parameter space to have a cyclic evolution, with the associated geometric phase α. For these reasons, AA regarded α as a geometric phase associated with a closed curve in the projective Hilbert space and not the parameter space, even in the special case considered by Berry. Given a cyclic evolution, an $H(t)$ that generates this evolution can be found so that the adiabatic approximation is valid. Then, α can be calculated with the use of the expression given by Berry in terms of the eigenstates of this Hamiltonian.

1.2.5 Example I: Born-Oppenheimer Approximation

The Born-Oppenheimer (BO) approximation is ubiquitous in quantum chemical calculations of molecular wavefunctions [26]. The approximation consists of two steps.

In the first step, the nuclear kinetic energy is neglected. This step is often justified by stating that the heavy nuclei move more slowly than the light electrons. Classically, this statement makes sense only if the momentum p of electrons and nuclei is of the same order of magnitude. In that case, $m_n \gg m_e$ implies $p^2/(2m_n) \ll p^2/(2m_e)$ (with m_n and m_e being the mass of the nucleus and the mass of the electron, respectively). It is straightforward to show that for two bodies in circular orbits around their center of mass (regardless of individual masses), the momentum of the two bodies is equal and opposite and that for any collection of particles in the center of mass frame, the net momentum is zero. Given that the center of mass frame is the lab frame (where the molecule is stationary), the momentum of the nuclei must be equal and opposite to that of the electrons. Recall that the corresponding operators do not contain mass, and consider the molecule to be a box containing the electrons and nuclei (as in the particle-in-a-box idealization). Since the kinetic energy is $p^2/(2m)$, it follows that the kinetic energy of the nuclei in a molecule is indeed typically much smaller than the kinetic energy of the electrons, the mass ratio being on the order of 10^4. Thus, the corresponding operator $\hat{T}_n$ is subtracted from the total molecular Hamiltonian. In the remaining electronic Hamiltonian $\hat{H}_e$, the nuclear positions enter as parameters. The electron-nucleus interactions are not removed, and the electrons still "feel" the Coulomb potential of the nuclei clamped at certain positions in space. This first step of the BO approximation is therefore often referred to as the clamped nuclei approximation.

The electronic Schrödinger equation

$$\hat{H}_e(\boldsymbol{r}, \boldsymbol{R})\chi(\boldsymbol{r}, \boldsymbol{R}) = E_e\chi(\boldsymbol{r}, \boldsymbol{R}), \tag{1.115}$$

is solved (out of necessity, it is solved only approximately). The quantity $\boldsymbol{r}$ corresponds to all electronic coordinates and $\boldsymbol{R}$ to all nuclear coordinates. The electronic energy eigenvalue E_e depends on the chosen positions $\boldsymbol{R}$ of the nuclei. Varying these positions $\boldsymbol{R}$ in small steps and repeatedly solving the electronic Schrödinger equation, one obtains E_e as a function of $\boldsymbol{R}$. This is the potential energy surface (PES): $E_e(\boldsymbol{R})$. Because this procedure of recomputing the electronic wave functions $\chi(\boldsymbol{r}, \boldsymbol{R})$ as a function of an infinitesimally changing nuclear geometry is reminiscent of the conditions for the adiabatic theorem, this manner of obtaining a PES is often referred to as the adiabatic approximation, and the PES itself is called an adiabatic surface. It is assumed, in accordance with the adiabatic theorem, that the same electronic state (the electronic ground state, for instance) is obtained upon small changes of the nuclear geometry. The method predicts a discontinuity (jump) in the PES if electronic state-switching occurs.

In the second step of the BO approximation, the nuclear kinetic energy $\hat{T}_n$ (containing partial derivatives with respect to the components of $\boldsymbol{R}$) is reintroduced; in addition, the Schrödinger equation for the nuclear motion is time independent, and the stationary wavefunctions $\phi(\boldsymbol{R})$ for the nuclei are obtained. It is traditional to use the word "motion" in this context even though (classically) motion implies time dependence. We solve the equation

$$\left[\hat{T}_n + E_e(\boldsymbol{R})\right]\phi(\boldsymbol{R}) = E\phi(\boldsymbol{R}), \tag{1.116}$$

This second step of the BO approximation involves separation of vibrational, translational, and rotational motions, which can be achieved by application of the Eckart conditions. The eigenvalue E is the total energy of the molecule, including contributions from electrons, nuclear vibrations, and the overall rotation and translation of the molecule.

1.2.6 Example II: Aharonov-Bohm Effect

We consider a magnetic field consisting of a single line with flux Φ. For the positions $\boldsymbol{R}$ not on the flux line, the magnetic field is zero; however, there must be a vector potential $\boldsymbol{A}(\boldsymbol{R})$ that satisfies

$$\oint_C \boldsymbol{A}(\boldsymbol{R}) \cdot d\boldsymbol{R} = \Phi \tag{1.117}$$

for circuits C threaded by the flux line. Aharonov and Bohm [27] showed that such vector potentials in quantum mechanics have physical significance even though they correspond to zero field. We show here how their effect can be interpreted as a geometric phase change [18].

Consider the quantum system consisting of particles with charge q confined to a box situated at $\boldsymbol{R}$ and not penetrated by any flux line (see Fig. 1.3). In the absence of flux ($\boldsymbol{A} = 0$), the Hamiltonian for the particle depends on the position $\boldsymbol{r}$ and its conjugate momentum $\boldsymbol{p}$ [18], i.e.,

Fig. 1.3 Aharonov-Bohm effect in a box transported around a flux line

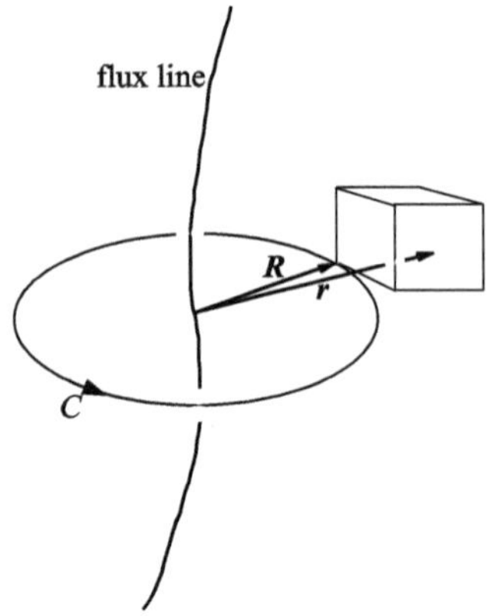

$$H = H(\boldsymbol{p}, \boldsymbol{r} - \boldsymbol{R}), \tag{1.118}$$

and the wavefunctions have the form $\psi_n(\boldsymbol{r} - \boldsymbol{R})$ with energies E_n independent of $\boldsymbol{R}$. With nonzero flux, the states $|n(\boldsymbol{R})\rangle$ satisfy

$$H(\boldsymbol{p} - q\boldsymbol{A}(\boldsymbol{r}), \boldsymbol{r} - \boldsymbol{R})|n(\boldsymbol{R})\rangle = E_n|n(\boldsymbol{R})\rangle, \tag{1.119}$$

an equation whose exact solutions can be obtained by multiplying ψ_n by an appropriate Dirac phase factor, giving

$$\langle \boldsymbol{r}|n(\boldsymbol{R})\rangle = \mathrm{e}^{i\frac{q}{\hbar}\int_{\boldsymbol{R}}^{\boldsymbol{r}} d\boldsymbol{r}'\cdot\boldsymbol{A}(\boldsymbol{r}')}\psi_n(\boldsymbol{r} - \boldsymbol{R}). \tag{1.120}$$

These solutions are single-valued in $\boldsymbol{r}$ and (locally) in $\boldsymbol{R}$. The energies are unaffected by the vector potential.

Let the box be transported around a circuit C threaded by the flux line. In this special case, it is not necessary to consider the transport adiabatic. After completion of the circuit, a geometric phase change is accumulated, which can be calculated from

$$\begin{aligned}\langle n(\boldsymbol{R})|\nabla_{\boldsymbol{R}} n(\boldsymbol{R})\rangle &= \int\int\int d^3\boldsymbol{r}\psi_n^*(\boldsymbol{r} - \boldsymbol{R})\left[-i\frac{q}{\hbar}\boldsymbol{A}(\boldsymbol{R})\psi_n(\boldsymbol{r} - \boldsymbol{R}) + \nabla_{\boldsymbol{R}}\psi_n(\boldsymbol{r} - \boldsymbol{R})\right] \\ &= -i\frac{q\boldsymbol{A}(\boldsymbol{R})}{\hbar}.\end{aligned} \tag{1.121}$$

The normalization of ψ_n leads to the vanishing of the second term in braces. Evidently, in this example, the analogy between $\mathrm{Im}\langle n|\nabla|n\rangle$ and a magnetic vector potential becomes a reality. As a result, one has

$$\lambda_n(C) = \frac{q}{\hbar}\oint_C \boldsymbol{A}(\boldsymbol{R})\cdot d\boldsymbol{R} = \frac{q\Phi}{\hbar}. \tag{1.122}$$

It is shown that the phase factor is independent of n and of C if this winds once around the flux line. The phase factor can be observed based on interference between the particles in the transported box and those in a box that is not transported around the circuit.

1.2.7 Example III: Adiabatic Quantum Computing

In quantum mechanics, a quantum state $|\psi\rangle$ evolves according to the following Schrödinger equation:

$$i\frac{d}{dt}|\psi(t)\rangle = H(t)|\psi(t)\rangle. \tag{1.123}$$

The adiabatic theorem [28] tells us how to follow this evolution when the Hamiltonian $H(t)$ varies slowly enough. Here, one can consider a smooth single-parameter family of Hamiltonians $\tilde{H}(s)$ with $0 \leq s \leq 1$ and take [29]

$$H(t) = \tilde{H}(\frac{t}{T}). \tag{1.124}$$

Thus, T controls the rate at which $H(t)$ varies. The instantaneous eigenstates and eigenenergies of $\tilde{H}(s)$ are determined by

$$H(s)|\ell; s\rangle = E_\ell(s)|\ell; s\rangle, \tag{1.125}$$

with

$$E_0(s) \leq E_1(s) \leq \cdots \leq E_{N-1}(s), \tag{1.126}$$

where N is the dimension of the Hilbert space. Assume that $|\psi(0)\rangle$ is the ground state of $\tilde{H}(0)$, that is,

$$|\psi(0)\rangle = |\ell = 0; s = 0\rangle. \tag{1.127}$$

According to the adiabatic theorem, if the gap between the two lowest levels, i.e., $E_1(s) - E_0(s)$, is strictly greater than zero for all $0 \leq s \leq 1$, then one has

$$\lim_{T\to\infty} |\langle \ell = 0; s = 1|\psi(T)\rangle| = 1. \tag{1.128}$$

This condition implies that if T is large enough, then the existence of a nonzero gap ensures that $|\psi(t)\rangle$ obeying (1.123) remains close to the instantaneous ground state of $H(t)$ of the form (1.124) for all t from 0 to T. One can define the minimum gap by

$$\Delta_{min} = \min_{0\leq s\leq 1} (E_1(s) - E_0(s)). \tag{1.129}$$

A closer look at the adiabatic theorem tells us that if

$$T \gg \frac{\mathscr{E}}{\Delta_{min}^2}, \tag{1.130}$$

where

$$\mathscr{E} = \max_{0 \le s \le 1} \left| \langle \ell = 1; s | \frac{d\tilde{H}}{ds} | \ell = 0; s \rangle \right|, \tag{1.131}$$

then one can make

$$|\langle \ell = 0; s = 1 | \psi(T) \rangle| \tag{1.132}$$

arbitrarily close to 1. For all of the problems that we discuss here, $\mathscr{E}$ is on the order of a typical eigenvalue of H and is not too large, so the size of T is determined by Δ_{min}^{-2}.

One can now apply adiabatic evolution to shift from the known ground state of H_B to the unknown ground state of H_P. Suppose that the ground state of H_P is unique. Consider

$$H(t) = (1 - \frac{t}{T}) \cdot H_B + \frac{t}{T} \cdot H_P. \tag{1.133}$$

From (1.124), one has

$$\tilde{H}(s) = (1 - s) \cdot H_B + s \cdot H_P. \tag{1.134}$$

Prepare the system so that it begins at $t = 0$ in the ground state of $H(0) = H_B$. According to the adiabatic theorem, if Δ_{min} is not zero and the system evolves according to (1.123), then for sufficiently large T, $|\psi(T)\rangle$ will be close to the ground state of H_P, that is, the solution to the computational problem [29].

1.2.8 Example IV: Geometric Quantum Computation

The physical realization of quantum computation requires a series of accurately controllable quantum gates. These gates can be implemented experimentally by using controlled geometric operations. It is remarkable that geometric operations based on adiabatic passages depend only on the global feature of the path executed [18] and thus provide a possible fault-tolerant way to perform quantum gate operations [30–33]. Similar to the case with adiabatic passage, the geometric gates based on nonadiabatic cyclic operations also depend only on some global features, making them robust to certain computational errors [25].

For a spin-$\frac{1}{2}$ particle subjected to an arbitrary magnetic field $\boldsymbol{B}$, the nonadiabatic cyclic geometric (or AA) phase [25] is just the solid angle determined by the evolution curve in the projective Hilbert space—a unit sphere S^2. Any two-component "spin" state $|\psi\rangle = [\mathrm{e}^{-i\phi/2}\cos(\theta/2), \mathrm{e}^{i\phi/2}\sin(\theta/2)]^T$ may be mapped into a unit vector $\boldsymbol{n} = (\sin\theta\cos\phi, \sin\theta\sin\phi, \cos\theta)$ in the projective Hilbert space via the relation

$\boldsymbol{n} = \langle\psi|\boldsymbol{\sigma}|\psi\rangle$, with the vector operator $\boldsymbol{\sigma}$ denoting the Pauli spin matrices. By varying the magnetic field, the AA phase is given by $\lambda = -\frac{1}{2}\int_C (1 - \cos\theta)d\phi$, where C lies along the actual evolution curve on S^2 and is determined by the equation $\partial_t \boldsymbol{n}(t) = -\mu \boldsymbol{B}(t) \times \boldsymbol{n}(t)/\hbar$. This phase recovers the Berry phase in adiabatic evolution [34].

For universal quantum computation, one need achieve only two kinds of noncommutable single-qubit gates and one nontrivial two-qubit gate [35]. The single-qubit Hamiltonian H is selected to pass through a cyclic evolution with period τ in the parameter space $\{\boldsymbol{B}\}$. One considers the process in which a pair of orthogonal states $|\psi_\pm\rangle$ can evolve cyclically. A phase difference between the two orthogonal states can be introduced by cyclically varying H. The phases acquired in this way contain both a geometric component and a dynamical component. The dynamical phase accumulated over the whole process can be removed by a simple method, and therefore, only the geometric phase needs to be considered. By taking into account the cyclic condition for $|\psi_\pm\rangle$ and removing the dynamical phase, one has $U(\tau)|\psi_\pm\rangle = \mathrm{e}^{\pm i\lambda}|\psi_\pm\rangle$, where $U(\tau)$ is the evolution operator. Here, we have also used the result that $\lambda[-\boldsymbol{n}(0)] = -\lambda[\boldsymbol{n}(0)]$ at any time if the two initial states correspond to $\pm\boldsymbol{n}(0)$ [34]. One can write an arbitrary initial state as $|\psi_i\rangle = a_+|\psi_+\rangle + a_-|\psi_-\rangle$, with $a_\pm = \langle\psi_\pm|\psi\rangle$, and express the two cyclic initial states as $|\psi_+\rangle = \cos\frac{\varphi}{2}|0\rangle + \sin\frac{\varphi}{2}|1\rangle$ and $|\psi_-\rangle = \cos\frac{\varphi}{2}|1\rangle - \sin\frac{\varphi}{2}|0\rangle$, where $|0\rangle$ and $|1\rangle$ constitute the computational basis for the qubit. The final state at time τ is found to be $|\psi_f\rangle = U(\varphi, \lambda)|\psi_i\rangle$, where

$$U(\varphi, \lambda) = \begin{pmatrix} \mathrm{e}^{i\lambda}\cos^2\frac{\varphi}{2} + \mathrm{e}^{-i\lambda}\sin^2\frac{\varphi}{2} & i\sin\varphi\sin\lambda \\ i\sin\varphi\sin\lambda & \mathrm{e}^{-i\lambda}\cos^2\frac{\varphi}{2} + \mathrm{e}^{i\lambda}\sin^2\frac{\varphi}{2} \end{pmatrix}. \tag{1.135}$$

It is straightforward to verify that the two operations $U^1(\varphi_1, \lambda_1)$ and $U^2(\varphi_2, \lambda_2)$ are noncommutable unless $\sin\lambda_1 \sin\lambda_2 \sin(\varphi_2 - \varphi_1) = 0$. Because two kinds of noncommutable operations constitute a universal set of single-bit gates, one can achieve the universal single-bit gates by choosing $\varphi_1 \neq \varphi_2 \pmod{2\pi}$ for any nontrivial phases λ_1 and λ_2. For example, the phase-flip gate $U_1 = \mathrm{e}^{-2\pi\lambda_1|1\rangle\langle 1|}$ (up to an irrelevant overall phase) can be achieved at $\varphi = 0$; the gate $U_2 = \mathrm{e}^{i\lambda_2\sigma_x}$ can be obtained at $\varphi = \pi/2$, which produces a spin flip (NOT operation) when $\lambda_2 = \pi/2$ and an equal-weight superposition of spin states when $\lambda_2 = \pi/4$.

In terms of the computational basis $\{|00\rangle, |01\rangle, |10\rangle, |11\rangle\}$, the unitary operator to describe the two-qubit gate is given by $U^{tq} = \mathrm{diag}[U(\lambda^0, \varphi^0), U(\lambda^1, \varphi^1)$ under the condition that the control qubit is far from the resonance condition for the operation of the target qubit. Here, $\lambda^\delta(\varphi^\delta)$ represents the geometric phase (the cyclic initial state) of the target qubit as long as the state of the control qubit corresponds to $\delta = 0, 1$ (where δ denotes the state of the control qubit). Following Ref. [35], one finds that the unitary operator U^{tq} is a nontrivial two-qubit gate if and only if $\lambda^1 \neq \lambda^0$ or $\varphi^1 \neq \varphi^0 \pmod{2\pi}$. Hence, all elements of quantum computation may be achievable by using nonadiabatic cyclic geometric operations [36].

1.2.9 Example V: Superadiabatic Quantum Driving

Quantum adiabatic processes are a powerful strategy to implement quantum state control, which aims at manipulating a quantum system to attain a prescribed target state in a controlled and optimal way [22, 37, 38]. However, often, such an adiabatic process may be too slow, and in nearly all adiabatic techniques, the population transfer is incomplete, with the fidelity close to, but less than, 1 [39]. Therefore, various protocols have been devised to speed up the process and to enhance the fidelity of quantum manipulation processes [40–45]. Superadiabatic (also known as transitionless or counterdiabatic) quantum driving [42, 43] is a valuable tool to speed up the adiabatic quantum behavior. The protocol suppresses the nonadiabatic transitions between energy eigenstates and ensures perfect adiabatic following by constructing an auxiliary field (Hamiltonian) [46].

The two-level system driven by an external coherent field is described by the dimensionless Schrödinger equation

$$i\frac{\partial}{\partial t}\begin{pmatrix} a \\ b \end{pmatrix} = H(t)\begin{pmatrix} a \\ b \end{pmatrix}, \tag{1.136}$$

with the Hamiltonian being

$$H(t) = \gamma(t)\sigma_z + v(t)\sigma_x, \tag{1.137}$$

where a and b are the probability amplitudes of diabatic states $|0\rangle$ and $|1\rangle$, respectively. The total probability $|a|^2 + |b|^2$ is conserved and set to be 1. σ_x and σ_z are Pauli matrices, and $\gamma(t)$ and $v(t)$ are the energy bias and coupling strength between the two diabatic levels, respectively.

The system has instantaneous adiabatic eigenstates $|\psi_\pm(t)\rangle$ that satisfy

$$H(t)|\psi_\pm(t)\rangle = \varepsilon_\pm(t)|\psi_\pm(t)\rangle, \tag{1.138}$$

where the eigenvalues $\varepsilon_\pm(t) = \pm\sqrt{\gamma^2 + v^2}$ and the subscripts $-$ and $+$ stand for the ground state and the excited state, respectively. Their difference $\varepsilon(t) = \varepsilon_+(t) - \varepsilon_-(t) = 2\sqrt{\gamma^2 + v^2}$ defines the energy splitting.

Assume that the system is initially prepared in the adiabatic ground state $|\psi_-(t_{ini})\rangle$ at time $t = t_{ini}$. The final state at time $t = t_{fin}$ is the state $|\psi_{fin}\rangle$ after an evolution of duration $t_{fin} - t_{ini}$. Our aim is to realize a superadiabatic protocol that ensures a perfect following of the instantaneous adiabatic ground state $|\psi_-(t)\rangle$ for all time. The protocol can drive the system from the starting state $|\psi_-(t_{ini})\rangle$ to the final state $|\psi_{fin}\rangle$ in an accelerated way and with high fidelity, i.e., the final state $|\psi_{fin}\rangle$ is as close as possible to the adiabatic ground state $|\psi_-(t_{fin})\rangle$, realizing a fidelity close to unity. Here, the fidelity function F_{fin} is defined as follows:

$$F_{fin} = |\langle\psi_{fin}|\psi_-(t_{fin})\rangle|^2, \tag{1.139}$$

which can be used to characterize the protocol efficiency.

The Hamiltonian (1.136) can be diagonalized using unitary transformation to a new basis (A, B), which is the adiabatic basis, given by

$$\begin{pmatrix} A \\ B \end{pmatrix} = U_0^{-1}(t) \begin{pmatrix} a \\ b \end{pmatrix}, \tag{1.140}$$

where U_0 is a rotation matrix that can be taken as

$$U_0 = \begin{pmatrix} -\sin\theta & \cos\theta \\ \cos\theta & \sin\theta \end{pmatrix}. \tag{1.141}$$

Here, the mixing angle $\theta = \frac{1}{2}\arctan(v(t)/\gamma(t))$. The Hamiltonian of the system in an adiabatic basis is

$$H'(t) = U_0^{-1} H(t) U_0 - i U_0^{-1} \dot{U}_0, \tag{1.142}$$

where the overdot represents the derivative with respect to time t. The first term is the diagonal part, while the second term is the non-diagonal part representing a nonadiabatic correction. The Schrödinger equation in the adiabatic basis reads

$$i\frac{\partial}{\partial t}\begin{pmatrix} A \\ B \end{pmatrix} = H'(t)\begin{pmatrix} A \\ B \end{pmatrix} = \begin{pmatrix} \varepsilon_- & -i\dot{\theta} \\ i\dot{\theta} & \varepsilon_+ \end{pmatrix}\begin{pmatrix} A \\ B \end{pmatrix}. \tag{1.143}$$

Adiabatic evolution occurs when the nonadiabatic coupling in the Hamiltonian is negligible relative to the eigenenergy splitting. Mathematically, adiabatic evolution requires the off-diagonal elements of the Hamiltonian (1.143) to be negligible relative to the diagonal ones, i.e. $|\dot{\theta}| \ll \varepsilon$, which expresses the adiabatic condition [42]. The efficiency of this transfer is limited by the adiabatic condition, which requires slow evolution. When the adiabatic condition cannot be fulfilled, a complete population transfer does not occur due to the effect of one or more nonadiabatic terms in the Hamiltonian. To overcome this challenge, one can construct an auxiliary Hamiltonian H_{cd} (also called a counter-diabatic field) that cancels the nonadiabatic part of the evolution under H alone [42, 43]. This process thus ensures a transitionless adiabatic following such that the system evolving under $H + H_{cd}$ always remains the instantaneous adiabatic ground state of H with probability 1, even for a finite duration of the protocol. In general, H_{cd} can be given by $H_{cd} = i\dot{U}_0 U_0^{-1}$. For a two-level system of the form described previously (1.136), one finds that [42, 43]

$$H_{cd}(t) = \frac{\partial\theta}{\partial t}\sigma_y, \tag{1.144}$$

where σ_y is the Pauli matrix. The total driving Hamiltonian is in the following form:

$$
\begin{aligned}
H_{tot}(t) &= H + H_{cd} \\
&= \gamma(t)\sigma_z + v(t)\sigma_x + \dot{\theta}\sigma_y \\
&= \begin{pmatrix} \gamma(t) & v(t) - i\dot{\theta} \\ v(t) + i\dot{\theta} & -\gamma(t) \end{pmatrix}.
\end{aligned} \tag{1.145}
$$

The Hamiltonian (1.145) can be described as a combination of an effective coupling and a phase term,

$$
H_{tot}(t) = \begin{pmatrix} \gamma(t) & v_{eff}(t)e^{-i\phi} \\ v_{eff}(t)e^{i\phi} & -\gamma(t) \end{pmatrix}, \tag{1.146}
$$

where $v_{eff}(t) = \sqrt{v^2(t) + \dot{\theta}^2}$. To eliminate the phase dependence, we apply the following transformation [47]:

$$
U_1 = \begin{pmatrix} e^{-i\phi/2} & 0 \\ 0 & e^{i\phi/2} \end{pmatrix}, \tag{1.147}
$$

which again provides a new set of basis functions; the resulting Hamiltonian is

$$
H_{tot}(t) = \begin{pmatrix} \gamma_{eff}(t) & v_{eff}(t) \\ v_{eff}(t) & -\gamma_{eff}(t) \end{pmatrix}, \tag{1.148}
$$

where $\gamma_{eff}(t) = \gamma(t) - \dot{\phi}/2$, with $\phi = \arctan(\dot{\theta}/v(t))$. This result implies that the effect of the extra field can also be achieved through an appropriate transformation $\gamma \to \gamma_{eff}$ and $v \to v_{eff}$.

Note that the shortcut to adiabaticity [44] is also a valuable tool to speed up the adiabatic quantum behavior. This method implements another reverse engineering approach using the Lewis-Riesenfeld (LR) invariant to carry the eigenstates of a Hamiltonian from a specified initial state to a final configuration and to then design the transient Hamiltonian from the LR invariant. Although different in form, these driving methods can be shown to be essentially equivalent to each other by properly adjusting the reference Hamiltonian [44]. Recently, these protocols have been extended to many quantum systems [47–54]. Experiments with superadiabatic protocols have been demonstrated for a Bose-Einstein condensate loaded into an accelerated optical lattice [55, 56], the electron spin of a single nitrogen vacancy center in diamond [57], a large single-photon detuning system with a cold atomic ensemble [58], and a continuous variable system for adiabatic transport of a trapped ion [59]. To illustrate this approach, some models have been adopted in these experiments, for instance, the Laudau-Zener model, the Allen-Eberly model, and the cosine model [55, 56, 58].

1.3 Classical-Quantum Correspondence

1.3.1 Bohr-Sommerfeld Quantization Rule

Planck introduced the constant (h or $\hbar = h/(2\pi)$) that now bears his name to eliminate the ultraviolet divergence in the black-body radiation spectrum. This constant defines a fundamental scale of action that allows us to relate kinematic and wave properties of quantum mechanical objects,

$$p = h/\lambda, \quad p = \hbar k; \tag{1.149}$$
$$E = h\nu, \quad E = \hbar\omega, \tag{1.150}$$

where the momentum p and the energy E describe the kinematic properties of the object, while the wavelength λ (or $k = 2\pi/\lambda$) and the frequency ν (or $\omega = 2\pi\nu$) describe the wave properties of the object. Quantum mechanical waves obey the superposition principle and may therefore exhibit quantum mechanical interference.

Experimentally, the emission and absorption spectra of atoms were known to exhibit discrete lines, which satisfied the Rydberg-Ritz combination principle, namely, $\nu = A_i - A_j$, where A_i and A_j are empirically determined terms. This discrete spectrum could not be explained by classical models, and one can now understand these terms as the energy levels of the atom. The arithmetic relationships among certain groups of lines were identified as early as 1885, when Balmer showed that certain lines were predicted by the formula $1/\lambda = R_H(1/2^2 - 1/n^2)$, with R_H being the Rydberg constant for hydrogen and $n = 3, 4, \ldots$. (Balmer, unaware of h, focused on the wavelength.) Similar series are associated with Lyman, Paschen, Brackett, and Pfund. Other experimental indications of quantization are as follows:

- *Franck-Hertz* Energy lost by electrons in inelastic collisions with atoms
- *Stern-Gerlach* Quantization of magnetic moment, spatial quantization
- *Photoelectric effect* Ultraviolet photons in, electrons out: photons as kinematic particles
- *Compton effect* Inelastic X-ray scattering: photon kinematics (Compton wavelength $\lambda_C \equiv h/mc \simeq 0.024\,\text{Å}$, with m being the electron rest mass and c being the speed of light)

Bohr proposed that the electrons in atoms can exist only in certain well-defined, stable orbits that satisfy the Bohr-Sommerfeld quantization condition,

$$\oint \boldsymbol{p} \cdot d\boldsymbol{q} = h(n + \frac{m}{4}), \quad n, m = 0, 1, 2, \ldots, \tag{1.151}$$

where $\boldsymbol{p}$ is the momentum and $\boldsymbol{q}$ is the position coordinate of an electron in three-dimensional space. The integral is performed over some closed orbit in phase space $\{\boldsymbol{p}, \boldsymbol{q}\}$; the integer n is the quantum number, and the integer m is the familiar Maslov index. Considering the electron as a wave with wavelength $\lambda = h/p$, this

Bohr-Sommerfeld quantization condition ensures that the wave is described by a function that is single-valued.

Although the Bohr atom is a crude approximation of the full solution to the Schrödinger equation in a central $1/r$ potential, this model does produce several correct results. Bohr required that his formulation of quantum mechanics satisfy the tested principles of classical mechanics in the limit of large quantum numbers, where the scale of action set by Planck's constant is small. This requirement is known as the correspondence principle.

1.3.2 Relation Between the Berry Phase and the Hannay Angle

In the previous sections, the adiabatic Berry phase and its geometrical meaning have been introduced from the perspective of quantum mechanics. Indeed, in classical mechanics, a geometrical angle exists that corresponds to the Berry phase. Based on both the adiabatic principle and Liouville's theorem, Hannay introduced the concept of a classical adiabatic angle in 1984 [3]. In the following year, Berry further discussed this classical angle (also called the "Hannay angle") and formulated its connection to the quantum adiabatic geometric phase [60].

Following Berry's derivations [60], we consider a quantum or classical system with N degrees of freedom whose Hamiltonian $H(\boldsymbol{p}, \boldsymbol{q}; \boldsymbol{R})$ depends on a set of slowly varying parameters $\boldsymbol{R} = (R_1, R_2, \ldots)$ in addition to the dynamical variables or operators $\boldsymbol{p} = (p_1, \ldots, p_N)$ and $\boldsymbol{q} = (q_1, \ldots, q_N)$. The evolution of the system is governed by the adiabatic theorem. In the quantum case [61], this theorem states that a system initially in an eigenstate, labeled by one or more parameters $n = \{n_i\}$, will remain in the same eigenstate $|n; \boldsymbol{R}(t)\rangle$, with energy $E_n(\boldsymbol{R}(t))$ as the $\boldsymbol{R}$ change. In the classical case [62], the theorem states that an orbit initially on an N-dimensional phase-space torus with actions $\boldsymbol{I} = \{I_j\}$ [63] will continue to explore the tori with the same values of $\boldsymbol{I}$ (adiabatic invariants), in spite of the varying Hamiltonian corresponding to $\boldsymbol{R}(t)$, provided that such tori continue to exist (for example, if the system remains integrable for all parameters $\boldsymbol{R}$).

These well-known adiabatic theorems fail to describe an important feature of the evolution, which manifests itself if the Hamiltonian returns to its original form after a (long) time T, i.e., $\boldsymbol{R}(T) = \boldsymbol{R}(0)$. One can describe such changes as taking the system around a circuit C in the space of parameters $\boldsymbol{R}$.

In quantum physics, the feature is a geometric phase factor $e^{i\lambda_n(C)}$ accumulated around C by a system in the nth state: if the state is initially $|\Psi(0)\rangle$, then the state at T is

$$|\Psi(T)\rangle = e^{i\lambda_n(C)} e^{-\frac{i}{\hbar}\int_0^T dt E_n(\boldsymbol{R}(t))} |\Psi(0)\rangle. \tag{1.152}$$

The second factor, which contains the familiar dynamical phase, is present even if the parameters remain constant, and the third factor $|\Psi(0)\rangle$ is an expression of the adiabatic theorem.

Classically, the feature that the adiabatic theorem does not describe is the shifts $\Delta\boldsymbol{\theta}(\boldsymbol{I}; C)$ in the angles $\boldsymbol{\theta} = \{\theta_j\}$ conjugate to the actions $\boldsymbol{I}$, in addition to those expected on the basis of the instantaneous frequencies $\boldsymbol{\omega} = \{\omega_j(\boldsymbol{I}; \boldsymbol{R})\}$: if the initial angles are $\boldsymbol{\theta}(0)$, then after the circuit C, the position of the system on its torus $\boldsymbol{I}$ is given by

$$\boldsymbol{\theta}(T) = \boldsymbol{\theta}(0) + \int_0^T dt \boldsymbol{\omega}(\boldsymbol{I}; \boldsymbol{R}(t)) + \Delta\boldsymbol{\theta}(\boldsymbol{I}; C). \tag{1.153}$$

The existence of $\Delta\boldsymbol{\theta}$ as a general feature of slowly cycled integrable systems was discovered by Hannay in 1984 [3].

The evolution of the angle variables, which by (1.153) determines the classical adiabatic angles $\Delta\boldsymbol{\theta}$, can be determined by making a canonical transformation to action-angle variables. This transformation can be achieved in terms of a generating function $S_\alpha(\boldsymbol{q}, \boldsymbol{I}; \boldsymbol{R}(t))$ according to the scheme [1]

$$(\boldsymbol{p}, \boldsymbol{q}) \leftarrow S_\alpha(\boldsymbol{q}, \boldsymbol{I}; \boldsymbol{R}(t)) \rightarrow (\boldsymbol{I}, \boldsymbol{\theta}), \; \boldsymbol{p}_\alpha = \frac{\partial S_\alpha}{\partial \boldsymbol{q}}, \; \boldsymbol{\theta}_\alpha = \frac{\partial S_\alpha}{\partial \boldsymbol{I}}. \tag{1.154}$$

In these formulae, the superscript α labels the branches of S, a function whose unavoidable multivaluedness reflects the fact that for a given torus I, q does not uniquely determine p.

The new Hamiltonian $\bar{H}(\boldsymbol{I}, \boldsymbol{\theta}; t)$ differs from the old Hamiltonian $H(\boldsymbol{p}, \boldsymbol{q}; \boldsymbol{R})$ in value and in functional form because the canonical transformation is time dependent through the slowly varying parameters $\boldsymbol{R}(t)$. In fact,

$$\bar{H}(\boldsymbol{I}, \boldsymbol{\theta}; t) = \mathscr{H}(\boldsymbol{I}; \boldsymbol{R}(t)) + \frac{\partial S_\alpha(\boldsymbol{q}, \boldsymbol{I}; \boldsymbol{R}(t))}{\partial \boldsymbol{R}} \cdot \frac{d\boldsymbol{R}}{dt}, \tag{1.155}$$

where

$$\mathscr{H}(\boldsymbol{I}; \boldsymbol{R}(t)) \equiv H(\boldsymbol{p}(\boldsymbol{I}, \boldsymbol{\theta}; \boldsymbol{R}(t)), \boldsymbol{q}(\boldsymbol{I}, \boldsymbol{\theta}; \boldsymbol{R}(t)); \boldsymbol{R}(t)) \tag{1.156}$$

is the (angle-independent) "action" Hamiltonian corresponding to constant $\boldsymbol{R}$. At any time t, the branch α and the value of q occurring in (1.155) are uniquely defined by $\boldsymbol{\theta}$ and $\boldsymbol{I}$.

To obtain an explicit form for $\bar{H}$, one can define the single-valued function

$$\mathscr{S}(\boldsymbol{I}, \boldsymbol{\theta}; \boldsymbol{R}) \equiv S_\alpha(\boldsymbol{q}(\boldsymbol{I}, \boldsymbol{\theta}; \boldsymbol{R}), \boldsymbol{I}; \boldsymbol{R}), \quad (0 \le \theta < 2\pi) \tag{1.157}$$

so that

$$\frac{\partial S_\alpha}{\partial \boldsymbol{R}} = \frac{\partial \mathscr{S}}{\partial \boldsymbol{R}} - \frac{\partial S_\alpha}{\partial \boldsymbol{q}} \cdot \frac{\partial \boldsymbol{q}}{\partial \boldsymbol{R}} = \frac{\partial \mathscr{S}}{\partial \boldsymbol{R}} - \boldsymbol{p}_\alpha \cdot \frac{\partial \boldsymbol{q}}{\partial \boldsymbol{R}}. \tag{1.158}$$

Finally, the new Hamiltonian reads

$$\bar{H}(\boldsymbol{I},\boldsymbol{\theta};t) = \mathscr{H}(\boldsymbol{I};\boldsymbol{R}(t)) + \frac{d\boldsymbol{R}}{dt}\cdot\left(\frac{\partial\mathscr{S}(\boldsymbol{I},\boldsymbol{\theta};\boldsymbol{R})}{\partial\boldsymbol{R}} - \boldsymbol{p}(\boldsymbol{I},\boldsymbol{\theta};\boldsymbol{R})\cdot\frac{\partial\boldsymbol{q}(\boldsymbol{I},\boldsymbol{\theta};\boldsymbol{R})}{\partial\boldsymbol{R}}\right). \tag{1.159}$$

This expression is globally single-valued because $\boldsymbol{q}$ and $\boldsymbol{p}$ are periodic functions of $\boldsymbol{\theta}$, and the increment of $\mathscr{S}$ around a circuit is

$$\mathscr{S}(\boldsymbol{\theta}+2\pi,\boldsymbol{I};\boldsymbol{R}) - \mathscr{S}(\boldsymbol{\theta},\boldsymbol{I};\boldsymbol{R}) = \oint \boldsymbol{p}d\boldsymbol{q} = 2\pi\boldsymbol{I}, \tag{1.160}$$

which does not depend on $\boldsymbol{R}$.

Hamilton's equation for the angle variables is

$$\dot{\theta}_j = \frac{\partial\bar{H}}{\partial I_j}, \quad j = 1,2,\ldots,N, \tag{1.161}$$

where the overdot denotes the time derivative. When applied to (1.159), the first term gives the part of the evolution that would occur even if the parameters remained constant, arising from the frequencies

$$\omega_j(\boldsymbol{I};\boldsymbol{R}) = \frac{\partial\mathscr{H}(\boldsymbol{I};\boldsymbol{R})}{\partial I_j}. \tag{1.162}$$

What we are seeking, however, is the angle shift defined by (1.153), and this parameter arises from the second term in (1.159):

$$\Delta\theta_j = \int_0^T dt\frac{d\boldsymbol{R}}{dt}\cdot\frac{\partial}{\partial I_j}\left(\frac{\mathscr{S}}{\partial\boldsymbol{R}} - \boldsymbol{p}\cdot\frac{\partial\boldsymbol{q}}{\partial\boldsymbol{R}}\right). \tag{1.163}$$

As it stands, this integral is challenging to evaluate because the integrand depends on time implicitly through the changes in $\boldsymbol{\theta}$ and $\boldsymbol{I}$ and explicitly through the variations of $\boldsymbol{R}$. It is natural at this point to invoke the adiabatic technique [63] of averaging over the implicit (fast) variations by integrating over the torus at each time t. We now give a brief introduction to this technique. Assume a classical system $H_0(\boldsymbol{I})$ subject to a perturbation of $\varepsilon H_1(\boldsymbol{I},\boldsymbol{\theta})$ (ε is a small quantity); the corresponding Hamilton's equations are

$$\dot{\boldsymbol{\theta}} = \boldsymbol{\omega}(\boldsymbol{I}) + \varepsilon\boldsymbol{f}(\boldsymbol{I},\boldsymbol{\theta}), \quad \dot{\boldsymbol{I}} = \varepsilon g(\boldsymbol{I},\boldsymbol{\theta}), \tag{1.164}$$

where

$$\boldsymbol{\omega}(\boldsymbol{I}) = \frac{\partial H_0}{\partial\boldsymbol{I}}, \quad \boldsymbol{f}(\boldsymbol{I},\boldsymbol{\theta}) = \frac{\partial H_1}{\partial\boldsymbol{I}}, \quad g(\boldsymbol{I},\boldsymbol{\theta}) = -\frac{\partial H_1}{\partial\boldsymbol{\theta}}. \tag{1.165}$$

According to the adiabatic averaging technique, one can replace Eq. (1.164) with another set, i.e.,

$$\dot{\boldsymbol{J}} = \varepsilon \bar{g}(\boldsymbol{J}), \;\; \bar{g}(\boldsymbol{J}) = \frac{1}{(2\pi)^N} \int_0^{2\pi} \cdots \int_0^{2\pi} g(\boldsymbol{J}, \boldsymbol{\theta}) d\theta_1 \cdots d\theta_N. \tag{1.166}$$

Arnold concluded that Eq. (1.166) are a suitable approximation for Eq. (1.164) [63]. The adiabatic averaging techniqueadiabatic averaging technique means that in the general case, the motion described by Eq. (1.164) includes two parts: the evolution part—as demonstrated in Eq. (1.166)—and the small oscillation part. For general equations, this argument is not accurate. However, for the canonical equations this method works well, and

$$\bar{g}(\boldsymbol{J}) = \frac{1}{(2\pi)^N} \int_0^{2\pi} \frac{\partial H_1(\boldsymbol{J}, \boldsymbol{\theta})}{\partial \boldsymbol{\theta}} \cdot d\boldsymbol{\theta} = 0. \tag{1.167}$$

When the above technique is applied to the Hamilton equation conjugate to (1.161), the actions $\boldsymbol{I}$ remain constant in spite of the (slow) variations in $\boldsymbol{R}$—which is of course the familiar adiabatic theorem. When applied to (1.97), one obtains

$$\Delta\theta_j(\boldsymbol{I}; C) = \oint d\boldsymbol{R} \cdot \frac{\partial}{\partial I_j} \frac{1}{(2\pi)^N} \oint d\boldsymbol{\theta} \left(\frac{\partial \mathscr{S}}{\partial \boldsymbol{R}} - \boldsymbol{p} \cdot \frac{\partial \boldsymbol{q}}{\partial \boldsymbol{R}} \right), \tag{1.168}$$

where $\oint d\boldsymbol{\theta} = \Pi_{j=1}^N \int_0^{2\pi} d\theta_j$. Equation (1.168) has the form of a line integral over a single-valued function in parameter space. The first term vanishes because $\partial \mathscr{S} / \partial \boldsymbol{R}$ is a gradient. The second term can be transformed by Stokes' theorem into an integral over any surface A in parameter space whose boundary is C. In the language of differential forms [63],

$$\Delta\theta_j(\boldsymbol{I}; C) = -\frac{\partial}{\partial I_j} \int\int_{A(C)} d\boldsymbol{A} \cdot \boldsymbol{W}(\boldsymbol{I}; \boldsymbol{R}), \tag{1.169}$$

where $d\boldsymbol{A}$ is the area element and $\boldsymbol{W}(\boldsymbol{I}; \boldsymbol{R})$ is the angle 2-form, which is given by

$$\boldsymbol{W}(\boldsymbol{I}; \boldsymbol{R}) = \frac{1}{(2\pi)^N} \oint d\boldsymbol{\theta} \left(\nabla_{\boldsymbol{R}} p_j(\boldsymbol{I}, \boldsymbol{\theta}; \boldsymbol{R}) \wedge \nabla_{\boldsymbol{R}} q_j(\boldsymbol{I}, \boldsymbol{\theta}; \boldsymbol{R}) \right). \tag{1.170}$$

The formulae (1.169) and (1.170) for the classical adiabatic angles constitute one version of the expressions obtained by Hannay [3]. It is straightforward to show that Hannay's angles are invariant under parameter-dependent and action-dependent deformations of the (arbitrary) origin from which the angles $\boldsymbol{\theta}$ are measured, i.e., under

$$\boldsymbol{\theta} \to \boldsymbol{\theta} + \boldsymbol{\beta}(\boldsymbol{I}; \boldsymbol{R}), \tag{1.171}$$

provided that $\boldsymbol{\beta}(\boldsymbol{I}; \boldsymbol{R})$ is single-valued across the area A in parameter space. This classical invariance corresponds to the invariance of the quantum phase factor under parameter-dependent changes in the phases of the eigenvectors $|n, \boldsymbol{R}\rangle \to \mathrm{e}^{i\chi(\boldsymbol{R})}|n, \boldsymbol{R}\rangle$ (see the appendix of [18]).

We now discuss the relation between the Hannay angle and the Berry phase. The quantum geometric phase $\lambda_n(C)$ defined by Eq. (1.152) is written in the form

$$\lambda_n(C) = -\int\int_{A(C)} d\boldsymbol{A} \cdot \boldsymbol{B}(n;\, \boldsymbol{R}), \tag{1.172}$$

where

$$\boldsymbol{B}(n;\, \boldsymbol{R}) = \mathrm{Im}\nabla_{\boldsymbol{R}} \wedge \langle n;\, \boldsymbol{R}|\nabla_{\boldsymbol{R}}|n;\, \boldsymbol{R}\rangle. \tag{1.173}$$

Clearly, Eqs. (1.172) and (1.173) are analogous to Eqs. (1.169) and (1.170), respectively. In position representation, one defines the wavefunction ψ_n by

$$\psi_n(\boldsymbol{q};\, \boldsymbol{R}) \equiv \langle \boldsymbol{q}|n;\, \boldsymbol{R}\rangle, \tag{1.174}$$

so the phase 2-form becomes

$$\boldsymbol{B}(n;\, \boldsymbol{R}) = \mathrm{Im}\nabla_{\boldsymbol{R}} \wedge \int d\boldsymbol{q}\psi_n^*(\boldsymbol{q};\, \boldsymbol{R})\nabla_{\boldsymbol{R}}\psi_n(\boldsymbol{q};\, \boldsymbol{R}), \tag{1.175}$$

where

$$\int d\boldsymbol{q} \equiv \Pi_{j=1}^{N} \int_{-\infty}^{+\infty} dq_j. \tag{1.176}$$

Semiclassically, ψ_n is associated with a torus whose actions are quantized by the corrected Bohr-Sommerfeld rule [64],

$$I_j = (n_j + m_j)\hbar, \tag{1.177}$$

where m_j are N constants whose values are unimportant in the present context. The wavefunction is obtained from the torus by projection from phase space to $\boldsymbol{q}$ space according to the method of Maslov (see [65] and as expressed in the simplified presentations by [66, 67]):

$$\psi_n(\boldsymbol{q};\, \boldsymbol{R}) = \sum a_\alpha(\boldsymbol{q}, \boldsymbol{I};\, \boldsymbol{R})\mathrm{e}^{\frac{i}{\hbar}S_\alpha(\boldsymbol{q}, \boldsymbol{I};\, \boldsymbol{R})}, \tag{1.178}$$

where S_α is the classical generating function, the summation over α corresponds to all branches p_α contributing at q, and the amplitudes are given in terms of the projection Jacobian by

$$a_\alpha^2 = \frac{1}{(2\pi)^N}\frac{d\boldsymbol{\theta}_\alpha}{d\boldsymbol{q}} = \frac{1}{(2\pi)^N}\det\left(\frac{d\theta_i}{dq_j}\right). \tag{1.179}$$

This quantity may be positive or negative, corresponding to $\pi/2$ phase shifts across turning points.

When the wavefunction (1.178) is substituted into (1.175), products of contributions from different branches α give rapid oscillations and cancel semiclassically upon integrating over $\boldsymbol{q}$, leaving

$$\boldsymbol{B}(n;\boldsymbol{R})=\frac{1}{\hbar}\nabla_{\boldsymbol{R}}\wedge\int d\boldsymbol{q}\frac{1}{(2\pi)^N}\sum_{\alpha}\frac{d\boldsymbol{\theta}_{\alpha}}{d\boldsymbol{q}}\nabla_{\boldsymbol{R}}S_{\alpha}(\boldsymbol{q},\boldsymbol{I};\boldsymbol{R}). \tag{1.180}$$

Transformation of the variables of integration from $\boldsymbol{q}$ to $\boldsymbol{\theta}$ and use of the formulae (1.157) and (1.158) give

$$\begin{aligned}\boldsymbol{B}(n;\boldsymbol{R}) &=\frac{1}{\hbar}\nabla_{\boldsymbol{R}}\wedge\frac{1}{(2\pi)^N}\oint d\boldsymbol{\theta}(\nabla_{\boldsymbol{R}}\mathscr{S}-\boldsymbol{p}\cdot\nabla_{\boldsymbol{R}}\boldsymbol{q})\\ &=-\frac{1}{\hbar(2\pi)^N}\oint d\boldsymbol{\theta}\nabla_{\boldsymbol{R}}p_j(\boldsymbol{I},\boldsymbol{\theta};\boldsymbol{R})\wedge\nabla_{\boldsymbol{R}}q_j(\boldsymbol{I},\boldsymbol{\theta};\boldsymbol{R})\\ &=-\frac{1}{\hbar}\boldsymbol{W}(\boldsymbol{I};\boldsymbol{R}),\end{aligned} \tag{1.181}$$

thus relating the phase 2-form to the angle 2-form (1.181).

Finally, this relation, together with formulae (1.169) and (1.172), immediately gives the connection between Hannay's angles and the geometric phase:

$$\Delta\theta_j(\boldsymbol{I};C)=-\hbar\frac{\partial}{\partial I_j}\lambda_n(C)=-\frac{\partial\lambda_n(C)}{\partial n_j}, \tag{1.182}$$

where the association (1.177) enables the quantum numbers n_j to be considered continuous variables.

1.3.3 Nonadiabatic Geometric Phase and Hannay Angle in the Generalized Harmonic Oscillator

In this section, we study the nonadiabatic geometric phase of the general cyclic evolutions of a generalized harmonic oscillator and demonstrate its relation with the Hannay angle [68]. To this end, an alternative method—the squeezed state approach—is introduced and applied.

The squeezed state approach has been successfully applied in many branches of physics, including quantum optics, high-energy physics, and condensed matter physics. The past several years have witnessed a growing application of the squeezed state to study chaotic dynamical systems [69–72]. In this section, we shall employ this approach to discuss the geometric phase and the Hannay angle for a generalized harmonic oscillator [68]. The reason for this choice is that this system admits the squeezed state as an exact solution. The squeezed state approach [73–75] starts from the time-dependent variational principle (TDVP) formulation,

$$\delta \int dt \langle \Phi, t | i\hbar \frac{\partial}{\partial t} - \hat{H} | \Phi, t \rangle = 0. \tag{1.183}$$

Variation with respect to $\langle \Phi, t|$ and $|\Phi, t\rangle$ gives rise to the Schrödinger equation and its complex conjugate, respectively. The squeezed state is chosen as the trial wave function, which is defined by the ordinary harmonic oscillator displacement operator acting on a squeezed vacuum state $|0\rangle$:

$$|\Psi\rangle = \mathrm{e}^{\alpha \hat{a}^\dagger - \alpha^* \hat{a}} |\phi\rangle, \quad |\phi\rangle = \mathrm{e}^{\frac{1}{2}(\beta \hat{a}^{\dagger 2} - \beta^* \hat{a}^2)} |0\rangle, \tag{1.184}$$

where $\hat{a}^\dagger$ and $\hat{a}$ are, respectively, boson creation and annihilation operators that satisfy the canonical commutation relation: $[\hat{a}, \hat{a}^\dagger] = 1$.

From the TDVP, we obtain the dynamical equations for the expectation values (q, p) and the quantum fluctuations $\Delta p^2 \equiv \langle \Psi, t | (\hat{p} - p)^2 | \Psi, t \rangle = \hbar(\frac{1}{4G} + 4\Pi^2 G)$ and $\Delta q^2 \equiv \langle \Psi, t | (\hat{q} - q)^2 | \Psi, t \rangle = \hbar G$,

$$\dot{q} = \frac{\partial H_{eff}}{\partial p}, \quad \dot{p} = -\frac{\partial H_{eff}}{\partial q}; \quad \hbar \dot{G} = \frac{\partial H_{eff}}{\partial \Pi}, \quad \hbar \dot{\Pi} = -\frac{\partial H_{eff}}{\partial G}, \tag{1.185}$$

where the overdot denotes the time derivative. The effective Hamiltonian H_{eff} is defined on the extended space (q, p, G, Π), taking the form $H_{eff} = \langle \Psi, t | \hat{H} | \Psi, t \rangle$.

The time-dependent variational principle leaves an ambiguity of a time-dependent phase $\lambda(t)$, which can be resolved with the aid of the Schrödinger equation,

$$\dot{\lambda}(t) = \langle \Psi, t | i \frac{\partial}{\partial t} | \Psi, t \rangle - \frac{1}{\hbar} \langle \Psi, t | \hat{H} | \Psi, t \rangle. \tag{1.186}$$

This phase is well defined for general nonadiabatic and noncyclic evolution of a squeezed state. It represents a phase change of the squeezed state during a time evolution. Clearly, the phase consists of two parts. The second part, which characterizes the time of evolution, is the dynamical phase and can be rewritten as

$$\lambda_D(t) = -\frac{1}{\hbar} \int_0^t H_{eff} dt. \tag{1.187}$$

The first part can be viewed as a difference of the total phase and the dynamical phase. We call this component the geometric phase since it is just the Aharonov-Anandan's phase for the case of cyclic evolution. From the expression of the squeezed state, the geometric phase is equal to

$$\lambda_G(t) = \int_0^t \left[\frac{1}{2\hbar} (p\dot{q} - q\dot{p}) - \dot{\Pi} G \right] dt. \tag{1.188}$$

It is clear that the evolution of both the expectation values (q, p) and the quantum fluctuations (G, Π) contribute to the geometric phase. The contribution from the former one is explicitly $\hbar$ dependent, while the contribution from the latter one is $\hbar$ independent. For the case of cyclic evolution of the squeezed state, the quantum phase is equal to the sum of the projection areas on the coordinates plane (q, p) and fluctuation plane (G, Π) swept out by a periodic orbit of the effective Hamiltonian.

The Hamiltonian of the generalized harmonic oscillator takes the form

$$\hat{H}(q, p, t) = \frac{1}{2}[a(t)\hat{q}^2 + b(t)\hat{p}^2 + c(t)(\hat{q}\hat{p} + \hat{p}\hat{q})], \tag{1.189}$$

where the real parameters $a(t)$, $b(t)$, and $c(t)$ are time-periodic functions with common period T. Our discussions are restricted to the elliptic case, namely, $a(t)b(t) > c^2(t)$.

Applying the squeezed state to this system, one obtains from (1.189) an effective Hamiltonian in the extended phase space $(q, p; G, \Pi)$,

$$H_{eff}(q, p; G, \Pi; t) = H_{cl}(q, p, t) + \hbar H_{fl}(G, \Pi, t), \tag{1.190}$$

where

$$H_{cl} = \frac{1}{2}[a(t)q^2 + b(t)p^2 + 2c(t)qp] \tag{1.191}$$

describes the motion of the expectation values;

$$H_{fl} = \frac{1}{2}[a(t)G + b(t)(\frac{1}{4G} + 4\Pi^2 G) + 4c(t)G\Pi] \tag{1.192}$$

depicts the evolution of the quantum fluctuations.

Starting from this effective Hamiltonian, it is straightforward to analyze the dynamical properties. The motions of both degrees of freedom are decoupled. In the fluctuation plane (G, Π), whole motions are restricted on the invariant tori except for a unique T-periodic solution denoted by $[G_p(t), \Pi_p(t)]$. The Hamiltonian H_{cl}, which describes the motions of the expectation values (q, p), is identical to the Hamiltonian of the classical version of the system (1.189). The point $(q = 0, p = 0)$ is clearly a fixed point. Other motions are quasiperiodic trajectories confined on the tori. Through a canonical transformation, $q = q(\bar{I}, \bar{\phi}, t)$ and $p = p(\bar{I}, \bar{\phi}, t)$, the Hamiltonian $H_{cl}(q, p, t)$ can be transformed to a new Hamiltonian $\bar{H}(\bar{I}, t)$ that does not contain the angle variable $\bar{\phi}$. Its solution is described by $\bar{I} = \bar{I}_0$; $\bar{\phi}(t) = \bar{\phi}_0 + \int_0^t (\partial \bar{H}(\bar{I}_0, t)/\partial \bar{I}_0)dt$. Because this canonical transformation is explicitly time dependent, the new Hamiltonian $\bar{H}$ differs from the old one H_{cl} both in value and in functional form. Here, we introduce a function A to measure the difference:

$$A(\bar{\phi}, \bar{I}, t) = \bar{H}(\bar{I}, t) - H_{cl}(\phi(\bar{\phi}, \bar{I}, t), I(\bar{\phi}, \bar{I}, t), t). \tag{1.193}$$

Therefore, the classical nonadiabatic Hannay angle is

$$\Theta_H = \left\langle \int_0^T \frac{\partial A}{\partial \bar{I}} dt \right\rangle_{\bar{\phi}_0}, \tag{1.194}$$

where the bracket denotes averaging around the invariant torus: $\langle \cdots \rangle = \frac{1}{2\pi}\int_0^{2\pi} \cdots d\bar{\phi}_0$.

We now turn to the quantum system (1.189). Since this system is a time-periodic Hamiltonian system, the Floquet theory applies. A unitary time evolution operator referring to one period T, the so-called Floquet operator $\hat{U}(T)$, is worthy of consideration. We can construct a state as a superposition of an infinite number of squeezed states,

$$|S_1\rangle = c\int_0^{2\pi} e^{(i/\hbar)\bar{I}_0\bar{\phi}_0}|\bar{I}_0, \bar{\phi}_0; G_0, \Pi_0\rangle d\bar{\phi}_0, \tag{1.195}$$

where $|\bar{I}_0, \bar{\phi}_0; G_0, \Pi_0\rangle$ represents a squeezed state centered at $q(\bar{I}_0, \bar{\phi}_0, t=0)$ and $p(\bar{I}_0, \bar{\phi}_0, t=0)$ with fluctuations G_0 and Π_0. The parameters G_0 and Π_0 are chosen on the unique periodic orbit $[G_0 = G_p(t=0), \Pi_0 = \Pi_p(t=0)]$; c is a normalization constant.

Consider the situation that $\hat{U}(mT)$ [or $\hat{U}^m(T)$] acts on the state $|S_1\rangle$,

$$\hat{U}(mT)|S_1\rangle = c\int_0^{2\pi} e^{(i/\hbar)\bar{I}_0\bar{\phi}_0}e^{i\lambda}|\bar{I}_0, \bar{\phi}_0 + \bar{\phi}^m; G_0, \Pi_0\rangle d\bar{\phi}_0, \tag{1.196}$$

where $\bar{\phi}^m = \int_0^{mT}(\partial\bar{H}(\bar{I}_0, t)/\partial\bar{I}_0)dt$ and $\lambda = \lambda_D(mT) + \lambda_G(mT)$. The dynamical part is $\lambda_D(mT) = -\frac{1}{\hbar}\int_0^{mT} H_{eff}dt$, and the geometric part is $\lambda_G(mT) = \frac{1}{\hbar}\int_0^{mT}\frac{1}{2}(p\dot{q} - q\dot{p})dt - \int_0^{mT}\dot{\Pi}_pG_pdt$. These parts can be expressed correspondingly as

$$\lambda_D(mT) = \langle\lambda_D(mT)\rangle_{\bar{\phi}_0} + \{\lambda_D(mT)\}(\bar{\phi}_0), \tag{1.197}$$

$$\lambda_G(mT) = \langle\lambda_G(mT)\rangle_{\bar{\phi}_0} + \{\lambda_G(mT)\}(\bar{\phi}_0), \tag{1.198}$$

where $\langle\cdots\rangle_{\bar{\phi}_0}$ denotes the average over $\bar{\phi}_0$ as in (1.194); $\{\cdots\}(\bar{\phi}_0)$ represents the terms relating to $\bar{\phi}_0$. Then,

$$\langle\lambda_G(mT)\rangle_{\bar{\phi}_0} = \frac{m}{\hbar}\left\langle\int_0^T [\frac{1}{2}(p\dot{q} - q\dot{p})]dt\right\rangle_{\bar{\phi}_0} - m\oint G_p d\Pi_p. \tag{1.199}$$

Making the variable transformation $\bar{\phi}_0' = \bar{\phi}_0 + \bar{\phi}^m$, we have

$$\hat{U}(mT)|S_1\rangle = c\cdot e^{i\lambda_m^1}\int_{\bar{\phi}^m}^{2\pi+\bar{\phi}^m} e^{(i/\hbar)\bar{I}_0\bar{\phi}_0'}e^{i\{\lambda_D(mT)\}(\bar{\phi}_0')+i\{\lambda_G(mT)\}(\bar{\phi}_0')}|\bar{I}_0, \bar{\phi}_0'; G_0, \Pi_0\rangle d\bar{\phi}_0', \tag{1.200}$$

where $\lambda_m^1 = m(\lambda_G^R + \lambda_D^R)$. The geometric part and the dynamical part take the following forms:

$$\lambda_G^R = \frac{1}{\hbar}\left\{\left\langle\int_0^T [\frac{1}{2}(p\dot{q} - q\dot{p})]dt\right\rangle_{\phi_0} - \bar{I}_0\int_0^T \frac{\partial \bar{H}}{\partial \bar{I}_0}dt\right\} - \oint G_p d\Pi_p, \quad (1.201)$$

$$\lambda_D^R = -\frac{1}{\hbar}\left\langle\int_0^T H_{eff}dt\right\rangle_{\phi_0}. \quad (1.202)$$

The integral in (1.200) can be written as $\int_0^{2\pi}\cdots + \int_{2\pi}^{2\pi+\bar{\phi}^m}\cdots - \int_0^{\bar{\phi}^m}\cdots$. The last two terms cancel each other if and only if $e^{(i/\hbar)\bar{I}_0 2\pi} = 1$, which gives rise to

$$\bar{I}_0 = n\hbar, \quad (1.203)$$

which is simply the quantization rule without Maslov-Morse correction.

The motion of the expectation values (q, p) confined on the invariant torus $\bar{I}_0$ is quasiperiodic. The ergodicity of the motion guarantees that the temporal average is equivalent to the spatial average over a sufficiently long time. From the ergodicity principle, we can choose an integer r that is large enough that the phase change [see (1.197) and (1.198)] during the time interval rT does not relate to $\bar{\phi}_0$. We then construct a state $|S_r\rangle$ as in [68]:

$$|S_r\rangle = |S_1\rangle + \cdots + e^{-i\lambda_m^1}\hat{U}(mT)|S_1\rangle + \cdots + e^{-i\lambda_{r-1}^1}\hat{U}((r-1)T)|S_1\rangle, \quad (1.204)$$

and under condition (1.203), one can prove that [68]

$$\hat{U}(T)|S_r\rangle = e^{i(\lambda_D^R + \lambda_G^R)}|S_r\rangle. \quad (1.205)$$

In fact, the above relation indicates that the state $|S_r\rangle$ is an eigenstate of the Floquet operator and that n is the state number. The parameters λ_D^R and λ_G^R are, respectively, the dynamical phase and the geometric phase relating to the cyclic states.

To interpret the geometric phase λ_G^R expressed by (1.201), let us consider the following differential 2-form that is preserved under the canonical transformation, i.e., $dp \wedge dq - dH \wedge dt = d\bar{I} \wedge d\bar{\phi} - d\bar{H} \wedge dt$. We rewrite this expression in another form:

$$dp \wedge dq - d\bar{I} \wedge d\bar{\phi} = -d(\bar{H} - H_{cl}) \wedge dt. \quad (1.206)$$

Let us first integrate the above equation over one period T and then average over the variable $\bar{\phi}_0$. Keeping in mind the area meaning of the differential 2-form, one finds immediately that the term bracketed in the expression of the geometric phase (1.201) corresponds to the left-hand side of this equation, whereas the right-hand side is equal to $n\hbar$ times the classical Hannay angle [see (1.194)]. The $\frac{1}{2}$ relation between the last term in (1.201) and the classical angle is given by Ge and Child [76] and verified by our explicit perturbative results as follows. We can then obtain a simple relation between the geometric phase and the nonadiabatic Hannay angle:

$$\lambda_G^R = -\left(n + \frac{1}{2}\right)\Theta_H. \tag{1.207}$$

We now take a specific choice of the periodic parameters as an example to demonstrate the above approach and verify our findings. Set $a(t) = 1 + \varepsilon \cos(\omega t)$, $b(t) = 1 - \varepsilon \cos(\omega t)$, and $c(t) = \varepsilon \sin(\omega t)$. Our discussions are restricted to the elliptic case, namely, $a(t)b(t) > c^2(t)$, i.e., $\varepsilon < 1$. The perturbation method is employed in the following discussions. Our solutions of power series are accurate to second order.

We rewrite the classical Hamiltonian in terms of the action-angle variables, i.e., $q = \sqrt{2I}\sin\phi$ and $p = \sqrt{2I}\cos\phi$,

$$H_{cl} = H_0(I) + \varepsilon H_1(I, \phi), \tag{1.208}$$

where $H_0 = I$ and $H_1 = -I\cos(\omega t + 2\phi)$. It is convenient to employ the Lie transformation [2] method to make a canonical transformation so that the new Hamiltonian $\bar{H}(\bar{I})$ contains the action variable only:

$$\bar{H}(\bar{I}) = \bar{I} - \frac{\bar{I}}{\omega + 2}\varepsilon^2. \tag{1.209}$$

The generating functions are $w_1 = \frac{I\sin(\omega t + 2\phi)}{\omega + 2}$ and $w_2 = 0$. The relation between the old variables and the new variables is given by $(\phi, I) = \mathscr{I}^{-1}(\bar{\phi}, \bar{I})$, where the transformation operator $\mathscr{I}^{-1} = 1 + \varepsilon\mathscr{L}_1 + \varepsilon^2(\mathscr{L}_2/2 + \mathscr{L}_1^2/2)$. The operator $\mathscr{L}_n$ is the Lie operator defined by $\mathscr{L}_n = [w_n, \ ]$, with $[\ ,\ ]$ representing the Poisson bracket. Applying (1.194), (1.208), and (1.209), we arrive at the analytical expression of the classical angle,

$$\Theta_H = \frac{2\pi\varepsilon^2}{(\omega + 2)^2}. \tag{1.210}$$

Clearly, this classical nonadiabatic Hannay angle is independent of the action.

A T-periodic solution $[G_p(t), \Pi_p(t)]$ of the Hamiltonian H_{fl} can be derived by using the power-series expansion

$$G_p(t) = \frac{1}{2} - \frac{\cos(\omega t)}{\omega + 2}\varepsilon, \quad \Pi_p(t) = -\frac{\sin(\omega t)}{\omega + 2}\varepsilon. \tag{1.211}$$

Note that an arbitrary ω can be approached by a series of rational numbers such as q/p; we can repeat the above process by constructing a state as in (1.204), where $r = q$ [68]. Finally, we obtain the analytic expression of the geometric phase,

$$\lambda_G^R = -\left(\frac{\bar{I}_0}{\hbar} + \frac{1}{2}\right)\frac{2\pi\varepsilon^2}{(\omega + 2)^2}. \tag{1.212}$$

Considering the quantization rule $\bar{I}_0 = n\hbar$ and the explicit expression of Hannay's angle (1.210), the above equation coincides with relation (1.207).

The quantum phases are obtained explicitly and found to be $-(n + 1/2)$ times the Hannay angle. In the adiabatic limit, our $(n + 1/2)$ relation is identical to the elegant formula of Berry [60]. However, the semiclassical approximation has not been invoked. Furthermore, the quantum phase can be interpreted as a sum of the area difference on the expectation value plane through a canonical transformation and the area on the quantum fluctuation plane swept out by a periodic orbit. This interpretation gives a unified picture of the geometric meaning of the quantal phase for the adiabatic and nonadiabatic cases [68].

References

1. Landau, L.D., Lifshitz, E.M.: Mechanics. Course of Theoretical Physics, vol. 1, 3rd edn. Pergamon Press, New York (1976)
2. Lichtenberg, A.J., Lieberman, M.A.: Regular and Chaotic Dynamics, 2nd edn. Springer (1992)
3. Hannay, J.H.: J. Phys. A Math. Gen. **18**, 221 (1985)
4. Zhou, J.L.: From Ordered to Chaotic Motion in Celestial Mechanics. In: Sun, Y.S., Zhou, L.Y. (eds.) Lectures on the Fundamentals of Celestial Mechanics. World Scientific (2016)
5. Khein, A., Nelson, D.F.: Am. J. Phys. **61**, 170 (1993)
6. Schwinger, J.: Phys. Rev. **51**, 648 (1937)
7. Born, M., Fock, V.: Z. Phys. **51**, 165 (1928)
8. Ehrenfest, P.: Ann. Phys. (Berlin) **51**, 327 (1916); Zener, C.: Proc. R. Soc. A **137**, 696 (1932); Kato, T.: J. Phys. Soc. Jpn. **5**, 435 (1950); Gell-Mann, M., Low, F.: Phys. Rev. **84**, 350 (1951); Farhi, E., et al.: Science **292**, 472 (2001); Jansen, S., Ruskai, M.B., Seiler, R.: J. Math. Phys. **48**, 102111 (2007); ReiChardt, B.: Proceedings of 36th STOC (2004); Jaksic, V.: J. Math. Phys. **43**, 2807 (1993); Martinez, A.: J. Math. Phys. **35**, 3889 (1994); Liu, J., Wu, B., Niu, Q.: Phys. Rev. Lett. **90**, 170404 (2003)
9. Bohm, A., Mostafazadeh, A., Koizumi, H., Niu, Q., Zwanziger, J.: The Geometric Phase in Quantum Systems. Springer (2003)
10. Teufel, S.: Adiabatic Perturbation Theory in Quantum Dynamics. Springer (2003)
11. Chruściński, D., Jamiolkowski, A.: Geometric Phases in Classical and Quantum Mechanics. Springer (2004)
12. Niu, Q., Chang, M.C., Wu, B., Xiao, D., Cheng, R.: Physical Effects of Geometric Phases. World Scientific (2017)
13. Anderson, M.H.: Science **269**, 198 (1995); Davis, K.B.: Phys. Rev. Lett. **75**, 3969 (1995); Bradley, C.C., et al.: Phys. Rev. Lett. **75**, 1687 (1995)
14. Nielsen, M.A., Chuang, I.L.: Quantum Computation and Quantum Information. Cambridge University Press, Cambridge (2000); Childs, A.M., et al.: Phys. Rev. A **65**, 012322 (2002); Fu, L.B.: Phys. Rev. Lett. **92**, 130404 (2004); Sarandy, M.S., Lidar, D.A. Phys. Rev. Lett. **95**, 250503 (2005)
15. Marzlin, K., Sanders, B.C.: Phys. Rev. Lett. **93**, 160408 (2004)
16. Tong, D.M., Singh, K., Kwek, L.C., Oh, C.H.: Phys. Rev. Lett. **95**, 110407 (2005)
17. Liu, J., Fu, L.B.: Phys. Lett. A **370**, 17 (2007)
18. Berry, M.V.: Proc. R. Soc. London A **392**, 45 (1984)
19. Manini, N., Pistolesi, F.: Phys. Rev. Lett. **85**, 3067 (2000); Hasegawa, Y., Loidl, R., Baron, M., Badurek, G., Rauch, H.: Phys. Rev. Lett. **87**, 070401 (2001); Filipp, S., Sjöqvist, E.: Phys. Rev. Lett. **90**, 050403 (2003); Wong, H.M., Cheng, K.M., Chu, M.C.: Phys. Rev. Lett. **94**, 070406 (2005)

20. Liu, J., et al.: Phys. Rev. A **66**, 023404 (2002)
21. Garanin, D.A., Schilling, R.: Phys. Rev. B **66**, 174438 (2002); Hanns, A., et al.: Phys. Rev. B **62**, 13880 (2000)
22. Bergmann, K., Theuer, H., Shore, B.W.: Rev. Mod. Phys. **70**, 1003 (1998)
23. Wu, T.T., Yang, C.N.: Phys. Rev. D **12**, 3845 (1975)
24. Von Neumann, J., Wigner, E.P.: Phys. Z. **30**, 467 (1929)
25. Aharonov, Y., Anandan, J.: Phys. Rev. Lett. **58**, 1593 (1987)
26. Born, M., Oppenheimer, J.R.: Annalen der Physik (in German) **389**, 457 (1927)
27. Aharonov, Y., Bohm, D.: Phys. Rev. **115**, 485 (1959)
28. Messiah, A.: Quantum Mechanics, vol. II. North Holland, Wiley, Amsterdam, New York (1976)
29. Farhi, E., Goldstone, J., Gutmann, S., Sipser, M.: Quantum Computation by Adiabatic Evolution. quant-ph/0001106
30. Zanardi, P., Rasetti, M.: Phys. Lett. A **264**, 94 (1999)
31. Jones, J.A., Vedral, V., Ekert, A., Castagnoli, G.: Nature (London) **403**, 869 (2000)
32. Duan, L.M., Cirac, J.I., Zoller, P.: Science **292**, 1695 (2001)
33. Falci, G., Fazio, R., Palma, G.M., Siewert, J., Vedral, V.: Nature (London) **407**, 355 (2000)
34. Zhu, S.L., Wang, Z.D.: Phys. Rev. Lett. **85**, 1076 (2000)
35. Lloyd, S.: Phys. Rev. Lett. **75**, 346 (1995)
36. Zhu, S.L., Wang, Z.D.: Phys. Rev. Lett. **89**, 097902 (2002)
37. Král, P., Thanopulos, I., Shapiro, M.: Rev. Mod. Phys. **79**, 53 (2007)
38. Brif, C., Chakrabarti, R., Rabitz, H.: New J. Phys. **12**, 075008 (2010)
39. Torosov, B.T., Guérin, S., Vitanov, N.V.: Phys. Rev. Lett. **106**, 233001 (2011)
40. Emmanouilidou, A., Zhao, X.G., Ao, P., Niu, Q.: Phys. Rev. Lett. **85**, 1526 (2000)
41. Masuda, S., Nakamura, K.: Phys. Rev. A **84**, 043434 (2011); Carlini, A., Hosoya, A., Koike, T., Okudaira, Y.: Phys. Rev. Lett. **96**, 060503 (2006)
42. Demirplak, M., Rice, S.A.: J. Phys. Chem. A **107**, 9937 (2003); Demirplak, M., Rice, S.A.: J. Phys. Chem. B **109**, 6838 (2005); Demirplak, M., Rice, S.A.: J. Chem. Phys. **129**, 154111 (2008); Masuda, S., Rice, S.A.: J. Phys. Chem. C **119**, 14513 (2014)
43. Berry, M.V.: J. Phys. A Math. Theor. **42**, 365303 (2009); Lim, R., Berry, M.V.: J. Phys. A Math. Theor. **24**, 3255 (1991)
44. Torrontegui, E., Ibáñez, S.S., Martanez-Garaot, S., Modugno, M., del Campo, A., Guéry-Odelin, D., Ruschhaupt, A., Chen, X., Muga, J.G.: Adv. At. Mol. Opt. Phys. **62**, 117 (2013)
45. Dou, F.Q., Cao, H., Fu, L.B., Liu, J.: Phys. Rev. A **93**, 043419 (2016); Daems, D., Ruschhaupt, A., Sugny, D., Guérin, S.: Phys. Rev. Lett. **111**, 050404 (2013)
46. Dou, F.Q., Liu, J., Fu, L.B.: EPL **116**, 60014 (2016)
47. Paul, K., Sarma, A.K.: Phys. Rev. A **91**, 053406 (2015)
48. del Campo, A.: Phys. Rev. Lett. **111**, 100502 (2013); Campbell, S., De Chiara, G., Paternostro, M., Palma, G.M., Fazio, R.: Phys. Rev. Lett. **114**, 177206 (2015); Santos, A.C., Silva, R.D., Sarandy, M.S.: Phys. Rev. A **93**, 012311 (2016)
49. Lu, M., Xia, Y., Shen, L.T., Song, J., An, N.B.: Phys. Rev. A **89**, 012326 (2014); Takahashi, K.: Phys. Rev. E **87**, 062117 (2013)
50. Masuda, S., Güngördü, U., Chen, X., Ohmi, T., Nakahara, M.: Phys. Rev. A **93**, 013626 (2016); Santos, A.C., Silva, R.D., Sarandy, M.S.: Phys. Rev. A **93**, 012311 (2016)
51. Sun, Z., Zhou, L., Xiao, G., Poletti, D., Gong, J.: Phys. Rev. A **93**, 012121 (2016); Liang, Z.T., Yue, X., Lv, Q., Du, Y.X., Huang, W., Yan, H., Zhu, S.L.: Phys. Rev. A **93**, 040305 (2016)
52. Okuyama, M., Takahashi, K.: Phys. Rev. Lett. **117**, 070401 (2016)
53. Ibáñez, S., Martínez-Garaot, S., Chen, X., Torrontegui, E., Muga, J.G.: Phys. Rev. A **84**, 023415 (2012)
54. Schaff, J.F., Song, X.L., Vignolo, P., Capuzzi, P., Labeyrie, G.: Phys. Rev. A **82**, 033430 (2010)
55. Bason, M.G., Viteau, M., Malossi, N., Huillery, P., Arimondo, E., Ciampini, D., Fazio, R., Giovannetti, V., Mannella, R., Morsch, O.: Nat. Phys. **8**, 147 (2012)
56. Malossi, N., Bason, M.G., Viteau, M., Arimondo, E., Mannella, R., Morsch, O., Ciampini, D.: Phys. Rev. A **87**, 012116 (2013)

57. Zhang, J., Shim, J.H., Niemeyer, I., Taniguchi, T., Teraji, T., Abe, H., Onoda, S., Yamamoto, T., Ohshima, T., Isoya, J., Suter, D.: Phys. Rev. Lett. **110**, 240501 (2013)
58. Du, Y.X., Liang, Z.T., Li, Y.C., Yue, X.X., Lv, Q.X., Huang, W., Chen, X., Yan, H., Zhu, S.L.: Nat. Commun. **7**, 12479 (2016)
59. An, M., Lv, D.S., del Campo, A., Kim, K.: Nat. Commun. **7**, 12999 (2016)
60. Berry, M.V.: J. Phys. A Math. Gen. **18**, 15 (1985)
61. Messiah, A.: Quantum Mechanics. North-Holland, Amsterdam (1962)
62. Dirac, P.A.M.: Proc. R. Soc. **107**, 25 (1925)
63. Arnold, V.I.: Mathematical Methods of Classical Mechanics. Springer, New York (1978)
64. Keller, J.B.: Ann. Phys. (NY) **4**, 180 (1958)
65. Maslov, V.P., Fedoriuk, M.V.: Semiclassical Approximation in Quantum Mechanics. Reidel, Dordrecht (1981)
66. Percival, I.C.: Adu. Chem. Phys. **36**, 1 (1977)
67. Iooss, G., Helleman, R.H.G., Stora, R.: Chaotic Behavior of Deterministic Systems. North-Holland, Amsterdam (1983)
68. Liu, J., Hu, B., Li, B.: Phys. Rev. Lett. **81**, 1749 (1998); Liu, J., Hu, B., Li, B.: Phys. Rev. A **58**, 3448 (1998)
69. Pattanayak, A.K., Schieve, W.C.: Phys. Rev. Lett. **72**, 2855 (1994)
70. Zhang, W.M., Feng, D.H.: Phys. Rep. **252**, 1 (1995)
71. Liu, W.V., Schieve, W.C.: Phys. Rev. Lett. **78**, 3278 (1997)
72. Hu, B., Li, B., Liu, J., Zhou, J.L.: Phys. Rev. E **58**, 1743 (1998)
73. Jackiw, R., Kerman, A.: Phys. Lett. **71A**, 158 (1979)
74. Zhang, W.M., Feng, D.H., Gilmore, R.: Rev. Mod. Phys. **62**, 867 (1990)
75. Tsui, Y., Fujiwara, Y.: Prog. Theor. Phys. **86**, 443 (1991)
76. Ge, Y.C., Child, M.S.: Phys. Rev. Lett. **78**, 2507 (1997)

Chapter 2
Nonlinear Adiabatic Evolution of Quantum Systems

Abstract In this chapter, we discuss the physical origins of the nonlinearity in quantum many-body systems. Based on the nonlinear Schrödinger equation, we introduce the adiabatic evolution of the quantum states, including both eigenstates and noneigenstates, and we introduce the nonlinear geometric phase acquired by an eigenstate during the adiabatic evolution. A nonlinear two-mode model for Bose-Einstein condensates (BECs) is used to demonstrate the nonlinear adiabatic theory.

2.1 Physical Origins of Nonlinearity

The nonlinearity may be introduced as possible modifications of quantum mechanics on the fundamental level [1]. Here, we consider the physical origins of the nonlinearity that stems from a mean-field treatment of the interactions between atoms and other components.

2.1.1 Nonlinear Gross-Pitaevskii (GP) Equation

The experimental realization of BECs in dilute atomic gases provides a suitable opportunity to study quantum phenomena on a macroscopic scale. The investigation of BECs has become one of the most active areas of research in contemporary physics and draws on many different subfields of physics. Since the interactions between atoms play a key role in the behavior of ultracold atomic clouds, the concepts and methods from condensed matter physics are used extensively [2].

The GP equation [3] describes the zero-temperature properties of the nonuniform Bose gas when the scattering length a is much less than the mean interparticle spacing. One can derive the GP equation by treating the interaction between atoms in a mean-field approximation. The effective interaction between two atoms at low energy is $U_0 = 4\pi\hbar^2 a/m$, with m being the mass of the atom, which corresponds to a contact interaction, $U_0\delta(\boldsymbol{r} - \boldsymbol{r}')$, where $\boldsymbol{r}$ and $\boldsymbol{r}'$ are the positions of the two atoms. With the help of the Hartree or mean-field approach, in the fully condensed state,

J. Liu et al., *Nonlinear Adiabatic Evolution of Quantum Systems*,
https://doi.org/10.1007/978-981-13-2643-1_2

all bosons are in the same single-particle state $\phi(\boldsymbol{r})$, and one can express the wave function of the N-particle system as

$$\Psi(\boldsymbol{r}_1, \boldsymbol{r}_2, \ldots, \boldsymbol{r}_N) = \prod_{i=1}^{N} \phi(\boldsymbol{r}_i). \tag{2.1}$$

The normalized condition is given by $\int d\boldsymbol{r} |\phi(\boldsymbol{r})|^2 = 1$. In the mean-field treatment, the effective Hamiltonian reads

$$H = \sum_{i=1}^{N} \left[\frac{\boldsymbol{p}_i^2}{2m} + V(\boldsymbol{r}_i) \right] + U_0 \sum_{i<j} \delta(\boldsymbol{r}_i - \boldsymbol{r}_j), \tag{2.2}$$

where $\boldsymbol{p}_i$ is the momentum of the ith particle and $V(\boldsymbol{r}_i)$ denotes the external potential. The energy of the state (2.1) is given by

$$E = N \int d\boldsymbol{r} \left[\frac{\hbar^2}{2m} |\nabla\phi(\boldsymbol{r})|^2 + V(\boldsymbol{r})|\phi(\boldsymbol{r})|^2 + \frac{(N-1)}{2} U_0 |\phi(\boldsymbol{r})|^4 \right]. \tag{2.3}$$

Consider the uniform Bose gas of volume V; the wave function of an atom in the ground state is $V^{-1/2}$, and thus, the interaction energy of a pair of atoms is U_0/V. The energy of a state with N bosons all in the same state can be approximated by

$$E = \frac{N(N-1)}{2V} U_0 \approx \frac{1}{2V} n^2 U_0, \tag{2.4}$$

where $n = N/V$. Here, we have assumed that $N \gg 1$. For convenience, one can introduce the concept of the wave function of the condensed state

$$\psi(\boldsymbol{r}) = N^{1/2} \phi(\boldsymbol{r}). \tag{2.5}$$

The density of the atoms is given by $n(\boldsymbol{r}) = |\psi(\boldsymbol{r})|^2$. By neglecting the terms of order $1/N$, the energy of the system can be written as

$$E(\psi) = \int d\boldsymbol{r} \left[\frac{\hbar^2}{2m} |\nabla(\boldsymbol{r})|^2 + V(\boldsymbol{r})|\psi(\boldsymbol{r})|^2 + \frac{1}{2} U_0 |\psi(\boldsymbol{r})|^4 \right]. \tag{2.6}$$

To find the optimal form of ψ, one can minimize the energy (2.6) with respect to the variations of $\psi(\boldsymbol{r})$ and its complex conjugate $\psi^*(\boldsymbol{r})$ subject to the condition that $N = \int d\boldsymbol{r} |\psi(\boldsymbol{r})|^2$ be constant. One writes $\delta E - \mu \delta N = 0$, where the chemical potential μ is the Lagrange multiplier that ensures constancy of the particle number, and the variations of ψ and ψ^* may thus be taken as arbitrary. This procedure is equivalent to minimizing the quantity $E - \mu N$ at fixed μ. Equating to zero the variation of $E - \mu N$ with respect to $\psi^*(\boldsymbol{r})$ gives

$$-\frac{\hbar^2}{2m}\nabla^2\psi(\boldsymbol{r}) + V(\boldsymbol{r})\psi(\boldsymbol{r}) + U_0|\psi(\boldsymbol{r})|^2\psi(\boldsymbol{r}) = \mu\psi(\boldsymbol{r}), \tag{2.7}$$

which is the time-independent GP equation. Note that the eigenvalue is the chemical potential, not the energy per particle as it is for the usual (linear) Schrödinger equation. For noninteracting particles all in the same state, the chemical potential is equal to the energy per particle, but for interacting particles, it is not. For a uniform Bose gas, the GP equation (2.7) becomes $\mu = U_0|\psi(\boldsymbol{r})|^2 = U_0 n$, which can also be obtained by using the thermodynamic relation $\mu = \partial E/\partial N$ to (2.4).

To treat dynamical problems, one can adopt the time-dependent GP equation as follows:

$$i\hbar\frac{\partial\psi(\boldsymbol{r},t)}{\partial t} = -\frac{\hbar^2}{2m}\nabla^2\psi(\boldsymbol{r},t) + V(\boldsymbol{r})\psi(\boldsymbol{r},t) + U_0|\psi(\boldsymbol{r},t)|^2\psi(\boldsymbol{r},t), \tag{2.8}$$

which is the basis for the discussion of the dynamics of the condensate. The time-dependent GP equation (2.8) can also be derived from the action principle

$$\delta\int_{t_1}^{t_2} L dt = 0, \tag{2.9}$$

where the Lagrangian L is given by

$$\begin{aligned} L &= \int d\boldsymbol{r}\frac{i\hbar}{2}\left(\psi^*\frac{\partial\psi}{\partial t} - \psi\frac{\partial\psi^*}{\partial t}\right) - E \\ &= \int d\boldsymbol{r}\left[\frac{i\hbar}{2}\left(\psi^*\frac{\partial\psi}{\partial t} - \psi\frac{\partial\psi^*}{\partial t}\right) - \mathscr{E}\right]. \end{aligned} \tag{2.10}$$

Here, E is the energy (2.6), and the energy density is given by

$$\mathscr{E} = \frac{\hbar^2}{2m}|\nabla\psi|^2 + V(\boldsymbol{r})|\psi|^2 + \frac{U_0}{2}|\psi|^4. \tag{2.11}$$

With a physically motivated choice of trial function for ψ, this variational principle provides the foundation for approximate solutions of dynamical problems. The physical content of the GP equation (2.8) can be revealed by reformulating the equation as a pair of hydrodynamic equations [2].

2.1.2 Nonlinear Optical Fibers

Optical fibers are made of isotropic silica; furthermore, all optical fibers are weakly birefringent. The phase velocity birefringence causes continuous changes of the polarization of a mode propagating along the fiber. Random deformations of the fiber cause the two so-called principal polarizations to propagate at different velocities. Environmental effects change both the delay and the orientation of the principal polarizations. A digital optical transmission pulse along the fiber has components along each of the two principal polarization modes. The polarization mode dispersion (PMD) pulls these components apart. Thus, even small PMD becomes significant as the bit rate is increased. However, the peak pulse energy also increases with the bit rate such that the Kerr nonlinearity can become important. The pulses that propagate along an isotropic fiber with (anomalous) dispersion propagate without spreading in conjunction with the Kerr nonlinearity. Such pulses are called solitons. The solitons propagating along a birefringent fiber with phase velocity birefringence undergo continuous polarization change but may still maintain their pulse envelope; these solitons are called Manakov solitons. The Kerr effect provides a binding force between the two pulse polarizations that opposes pulse separation caused by the PMD. This process can prevent pulse breakup if the PMD is not excessive. This property of the Manakov solitons prompts their use in high bit-rate optical communication [4].

In the presence of strong polarization mixing by birefringence, the pulse propagation along a fiber with anomalous dispersion and the Kerr effect obeys the Manakov equation; i.e., the cross coupling between the two polarizations u and v along two orthogonal vectors $\hat{\boldsymbol{e}}^{(1)}$ and $\hat{\boldsymbol{e}}^{(2)}$ is equal to the self-coupling [5]

$$i\frac{\partial u}{\partial t} = \frac{1}{2}\omega''\frac{\partial^2}{\partial x^2}u + \kappa(|u|^2 + |v|^2)u, \tag{2.12}$$

$$i\frac{\partial v}{\partial t} = \frac{1}{2}\omega''\frac{\partial^2}{\partial x^2}v + \kappa(|u|^2 + |v|^2)v. \tag{2.13}$$

The evolution in time is described to emphasize the correspondence with the nonlinear Schrödinger equation. Here, ω'' is the second derivative of the frequency with respect to the propagation constant, $\omega'' = d^2\omega/d\beta^2$, where ω'' is positive when the dispersion is anomalous; κ is the Manakov Kerr coefficient.

In introducing the PMD perturbation, we interpret (2.12) and (2.13) in a very specific way [4]. We consider a section of fiber that is shorter than the coherence length of the PMD and, a fortiori, shorter than the soliton period. Within this short section of fiber, the two polarizations are interpreted as expressed in terms of the principal eigenvectors $\hat{\boldsymbol{e}}^{(1)}$ and $\hat{\boldsymbol{e}}^{(2)}$ of the PMD. The field vector $\hat{\boldsymbol{\psi}}$ is $\hat{\boldsymbol{\psi}} = u\hat{\boldsymbol{e}}^{(1)} + v\hat{\boldsymbol{e}}^{(2)}$. Within this short distance, the PMD generates a perturbation term due to the group delay difference. This perturbation is introduced into the two equations

$$i\frac{\partial u}{\partial t} = \frac{1}{2}\omega''\frac{\partial^2}{\partial x^2}u + \kappa(|u|^2 + |v|^2)u + i\xi_u'\frac{\partial}{\partial x}u, \tag{2.14}$$

$$i\frac{\partial v}{\partial t} = \frac{1}{2}\omega''\frac{\partial^2}{\partial x^2}v + \kappa(|u|^2 + |v|^2)v + i\xi_v'\frac{\partial}{\partial x}v, \tag{2.15}$$

where ξ_u' and ξ_v' are the group velocity deviations of the two principal polarizations.

2.1.3 Nonlinear Atom-Molecule Conversion

Let us consider a bosonic atom-molecule conversion system. The model is schematically sketched in Fig. 2.1a. The initial state $|a\rangle$ (atomic state) and the intermediate state $|e\rangle$ (excited molecular state) are coupled by a pump laser with Rabi frequency Ω_2, while the state $|e\rangle$ and the target state $|g\rangle$ (ground molecular state) are coupled by a Stokes laser with Rabi frequency Ω_1. The frequencies of the applied lasers are expressed in terms of the single- and two-photon detunings Δ and δ. Without loss of generality, we assume that the Rabi frequencies $\Omega_{1,2}$ are real and positive. Under the two-photon resonance condition, i.e., $\delta = 0$, the Hamiltonian in second-quantized form reads [6, 7]

$$H_{am} = -\hbar\Delta\hat{\psi}_e^\dagger\hat{\psi}_e + \frac{\hbar}{2}(-\Omega_2\hat{\psi}_e^\dagger\hat{\psi}_a\hat{\psi}_a + \Omega_1\hat{\psi}_g^\dagger\hat{\psi}_e + \text{H.c.}), \tag{2.16}$$

where $\hat{\psi}_i$ and $\hat{\psi}_i^\dagger$ ($i = a, e, g$) are the annihilation and creation operators for state $|i\rangle$, respectively. Under the mean-field approximation, in which $\hat{\psi}_i$ and $\hat{\psi}_i^\dagger$ are replaced by c number ψ_i and ψ_i^*, the nonlinear Schrödinger equations (with $\hbar = 1$) are

$$i\dot{\psi}_a = -\Omega_2\psi_a^*\psi_e, \tag{2.17}$$

$$i\dot{\psi}_e = -\Delta\psi_e - \frac{\Omega_2}{2}\psi_a^2 + \frac{\Omega_1}{2}\psi_g, \tag{2.18}$$

$$i\dot{\psi}_g = \frac{\Omega_1}{2}\psi_e, \tag{2.19}$$

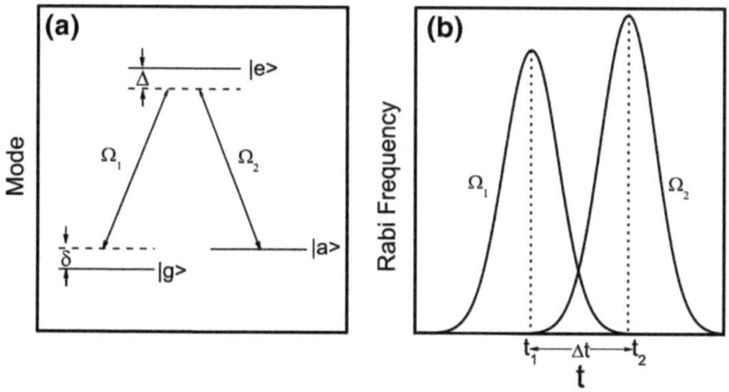

Fig. 2.1 **a** Three-level system coupled by two lasers. Ω_1 and Ω_2 are the Rabi frequencies for the pump and Stokes lasers, and Δ and δ are single- and two-photon detunings, respectively. **b** Time dependence of Ω_1 and Ω_2. t_1 and t_2 are the corresponding centers of the two pulses, and Δt is the time delay between the two pulses

where the overdot denotes the time derivative. In the above model, the nonlinear collisions between the particles are neglected, so nonlinearity arises solely because it takes two atoms to form one molecule. Mathematically, the Hamiltonian in the above Schrödinger equations is a function of the instantaneous wave function and its conjugate.

For the above Λ system, because the Hamiltonian is a function of both the wave function and its conjugate, the $U(1)$ invariance is broken. Instead, the system is invariant under the following transformation:

$$U(\phi) = \mathrm{e}^{i\boldsymbol{\Theta}(\phi)}, \quad \boldsymbol{\Theta}(\phi) = \begin{pmatrix} \phi & 0 & 0 \\ 0 & 2\phi & 0 \\ 0 & 0 & 2\phi \end{pmatrix}. \tag{2.20}$$

Under this transformation, $|\psi'\rangle = U(\phi)|\psi\rangle = (\psi_a \mathrm{e}^{i\phi}, \psi_e \mathrm{e}^{2i\phi}, \psi_g \mathrm{e}^{2i\phi})^T$. In fact, when the diagonal terms in the above matrix are identical, the transformation $U(\phi)$ degenerates to the $U(1)$ transformation.

2.2 Nonlinear Adiabatic Evolution of Quantum States

The adiabatic evolution has been an important method for the preparation and control of quantum states [8, 9]. The main guidance comes from the adiabatic theorem of quantum mechanics [10], which dictates that an initial nondegenerate eigenstate remains as an instantaneous eigenstate when the Hamiltonian changes slowly relative to the level spacings. More precisely, the quantum eigenstate evolves only in its phase, given by the time integral of the eigenenergy (dynamical phase) and a quantity independent of the time duration (geometric phase). The linearity of quantum mechanics then immediately allows a precise statement on the adiabatic evolution of noneigenstates through the superposition principle.

Our focus in this section is how the adiabatic theorem is modified in the nonlinear evolution of quantum states [11]. The challenges in the theoretical study of adiabatic control of the quantum states arise not only from the lack of unitarity (although the norm of a state is conserved, the inner product of two states changes with time) but also from the absence of the superposition principle [12, 13].

Here, we attempt to overcome these challenges by combining ideas from classical adiabatic dynamics and quantum geometric phases. Noting that the eigenstates correspond to the extremum points of the system energy, we find that their adiabatic condition depends on the Bogoliubov excitation spectrum about such points and is independent of the level spacings between the eigenstates. In addition, because of the nonlinearity, the adiabatic evolution of noneigenstates cannot be expressed as a superposition of such eigenstates with conserved probabilities as in the linear case [11]. We find that the Aharonov-Anandan phases [14], which can be defined for the cyclic or the quasicyclic quantum states (at fixed control parameters) can serve as the adiabatic invariants analogous to canonical actions in classical systems [15].

2.2.1 General Formalism

The Schrödinger equation, linear or nonlinear, can be written in the form (with $\hbar = 1$) [11]

$$i\frac{\partial}{\partial t}\Psi(\mathbf{r}, t) = H(\Psi, \Psi^*, \mathbf{r}, t)\Psi(\mathbf{r}, t). \tag{2.21}$$

By "nonlinear", we mean that the Hamiltonian H is not only a function of space $\mathbf{r}$ and time t but also explicitly depends on the wave function Ψ itself. For practical usage, one can typically separate the spatial and temporal variables by expanding the wave function over an orthonormal basis set $\varphi_k(\mathbf{r})$, i.e.,

$$\Psi(\mathbf{r}, t) = \sum_{k=1}^{N} \psi_k(t)\varphi_k(\mathbf{r}), \tag{2.22}$$

where $\psi_k(t)$ is the kth amplitude of the wave function $\Psi(\mathbf{r}, t)$. Inserting Eq. (2.22) into Eq. (2.21), multiplying the left- and right-hand sides by $\varphi_j^*(\mathbf{r})$ $(j = 1, \ldots, N)$ simultaneously, and then integrating both sides over $\mathbf{r}$, one obtains

$$i\frac{d\psi_j(t)}{dt} = \sum_{k=1}^{N} H_{jk}(\psi, \psi^*, t)\psi_k(t), \tag{2.23}$$

where $H_{jk}(\psi, \psi^*, t) = \int \varphi_j^*(\mathbf{r})H(\Psi, \Psi^*, \mathbf{r}, t)\varphi_k(\mathbf{r})d\mathbf{r}$. In matrix form, one obtains (for simplicity, we omit the variable t hereafter when not otherwise specified)

$$i\frac{d}{dt}\begin{pmatrix}\psi_1\\ \psi_2\\ \vdots\\ \psi_N\end{pmatrix} = \begin{bmatrix} H_{11} & H_{12} & \cdots & H_{1N}\\ H_{21} & H_{22} & \cdots & H_{2N}\\ \vdots & \vdots & \ddots & \vdots\\ H_{N1} & H_{N2} & \cdots & H_{NN}\end{bmatrix}\begin{pmatrix}\psi_1\\ \psi_2\\ \vdots\\ \psi_N\end{pmatrix}. \tag{2.24}$$

In the following, we are interested in the adiabatic evolution of a quantum system. We thus change the variable t to $\boldsymbol{R}$, which denotes all the system parameters subject to adiabatic change, and we rewrite the Schrödinger equation as

$$i\frac{d\psi_j}{dt} = \frac{\partial}{\partial \psi_j^*}\mathscr{H}(\psi, \psi^*, \boldsymbol{R}) \tag{2.25}$$

and its complex conjugate form as

$$i\frac{d\psi_j^*}{dt} = -\frac{\partial}{\partial \psi_j}\mathscr{H}(\psi, \psi^*, \boldsymbol{R}), \tag{2.26}$$

where $\mathscr{H}$ is the energy of the system. For the linear case, one obtains the simple relation $\mathscr{H} = \sum_{j=1}^{N} \sum_{k=1}^{N} \psi_j^*(t) H_{jk}(t) \psi_k(t) = \langle \psi | H | \psi \rangle$ from Eqs. (2.23) and (2.25). For the nonlinear case, however, H is a matrix that depends on the wave function, and thus, the above simple relation no longer holds.

The Schrödinger equation (2.25) and (2.26) has a canonical structure of classical dynamics. One can show, for example, that for each j, the probability $P_j = |\psi_j|^2$ and phase $Q_j = \arg(\psi_j)$ form a canonical pair that satisfies Hamilton's equations of motion with the energy $\mathscr{H}$ serving as the classical Hamiltonian. This result can be proven as follows:

$$\frac{dQ_j}{dt} = \frac{\partial Q_j}{\partial \psi_j} \frac{d\psi_j}{dt} + \frac{\partial Q_j}{\partial \psi_j^*} \frac{d\psi_j^*}{dt} = \frac{1}{2i\psi_j} \frac{d\psi_j}{dt} - \frac{1}{2i\psi_j^*} \frac{d\psi_j^*}{dt}, \tag{2.27}$$

$$\frac{\partial \mathscr{H}}{\partial P_j} = \frac{\partial \mathscr{H}}{\partial \psi_j} \frac{\partial \psi_j}{\partial P_j} + \frac{\partial \mathscr{H}}{\partial \psi_j^*} \frac{\partial \psi_j^*}{\partial P_j} = \frac{-i}{2\psi_j^*} \frac{d\psi_j^*}{dt} + \frac{i}{2\psi_j} \frac{d\psi_j}{dt}, \tag{2.28}$$

$$\frac{dP_j}{dt} = \psi_j \frac{d\psi_j^*}{dt} + \psi_j^* \frac{d\psi_j}{dt} = \psi_j \left(i \frac{\partial \mathscr{H}}{\partial \psi_j} \right) + \psi_j^* \left(-i \frac{\partial \mathscr{H}}{\partial \psi_j^*} \right), \tag{2.29}$$

$$\frac{\partial \mathscr{H}}{\partial Q_j} = \frac{\partial \mathscr{H}}{\partial \psi_j} \frac{\partial \psi_j}{\partial Q_j} + \frac{\partial \mathscr{H}}{\partial \psi_j^*} \frac{\partial \psi_j^*}{\partial Q_j} = i\psi_j \frac{\partial \mathscr{H}}{\partial \psi_j} - i\psi_j^* \frac{\partial \mathscr{H}}{\partial \psi_j^*}; \tag{2.30}$$

thus,

$$\frac{dQ_j}{dt} = -\frac{\partial \mathscr{H}}{\partial P_j}, \quad \frac{dP_j}{dt} = \frac{\partial \mathscr{H}}{\partial Q_j}, \quad (j = 1, \ldots, N). \tag{2.31}$$

In this proof, we have applied Eqs. (2.25) and (2.26) in addition to $\psi_j = \sqrt{P_j} e^{iQ_j}$ and $\psi_j^* = \sqrt{P_j} e^{-iQ_j}$, which lead to the relationships $\frac{\partial \psi_j}{\partial P_j} = \frac{1}{2\sqrt{P_j}} e^{iQ_j} = \frac{1}{2\psi_j^*}$, $\frac{\partial \psi_j^*}{\partial P_j} = \frac{1}{2\sqrt{P_j}} e^{-iQ_j} = \frac{1}{2\psi_j}$, $\frac{\partial \psi_j}{\partial Q_j} = i\sqrt{P_j} e^{iQ_j} = i\psi_j$, and $\frac{\partial \psi_j^*}{\partial Q_j} = -i\sqrt{P_j} e^{iQ_j} = -i\psi_j^*$, and, reversely, $Q_j = \frac{1}{2i} \ln \frac{\psi_j}{\psi_j^*}$, $\frac{\partial Q_j}{\partial \psi_j} = \frac{1}{2i} \frac{\psi_j^*}{\psi_j} \frac{1}{\psi_j^*} = \frac{1}{2i\psi_j}$, and $\frac{\partial Q_j}{\partial \psi_j^*} = \frac{1}{2i} \frac{\psi_j^*}{\psi_j} (-1) \frac{\psi_j}{(\psi_j^*)^2} = -\frac{1}{2i\psi_j^*}$. Due to Eq. (2.31), the nonlinear Schrödinger equation is often termed classical, although the linear Schrödinger equation has not been described in this way. In fact, one often regards the mean-field equation as the classical analog of the original many-body quantum problem [16]. However, we do not address this correspondence in the present section.

Our original system is quantum mechanical, which is expected to impose additional structures in the corresponding canonical dynamics. Specifically, the system has the gauge symmetry that $\mathscr{H}$ is invariant under a shift in the overall phase of the wave function. This condition implies that the total probability must be conserved and that the dynamics of the overall phase can be separated from the remaining degrees of freedom. For example, in a finite dimensional problem, we can choose a new set of variables: $p_j = |\psi_j|^2 = P_j, q_j = \arg(\psi_j) - \arg(\psi_N) = Q_j - Q_N, j = 1, 2, \ldots, N-1$, and $p_N = \sum_{j=1}^N |\psi_j|^2 = \sum_{j=1}^N P_j, q_N = \arg(\psi_N) = Q_N$. It is then straightforward to prove that

$$\{q_m, q_n\}_{P,Q} = \sum_{j=1}^{N} \left(\frac{\partial q_m}{\partial P_j} \frac{\partial q_n}{\partial Q_j} - \frac{\partial q_m}{\partial Q_j} \frac{\partial q_n}{\partial P_j} \right) = 0, \tag{2.32}$$

$$\{p_m, p_n\}_{P,Q} = \sum_{j=1}^{N} \left(\frac{\partial p_m}{\partial P_j} \frac{\partial p_n}{\partial Q_j} - \frac{\partial p_m}{\partial Q_j} \frac{\partial p_n}{\partial P_j} \right) = 0, \tag{2.33}$$

$$\{p_m, q_n\}_{P,Q} = \sum_{j=1}^{N} \left(\frac{\partial p_m}{\partial P_j} \frac{\partial q_n}{\partial Q_j} - \frac{\partial p_m}{\partial Q_j} \frac{\partial q_n}{\partial P_j} \right) = \delta_{mn}, \tag{2.34}$$

where $\{\cdot\}$ is the Poisson bracket, $m, n = 1, 2, \ldots, N$, which indicates that $P, Q \to p, q$ are canonical transformations. Because the Hamiltonian is independent of q_N, so p_N is conserved and can be set to unity, the other variables form a closed set of Hamiltonian dynamics:

$$\frac{dq_j}{dt} = \frac{\partial \mathscr{H}}{\partial p_j}, \quad \frac{dp_j}{dt} = -\frac{\partial \mathscr{H}}{\partial q_j}, \quad (j = 1, \ldots, N-1). \tag{2.35}$$

The gauge symmetry also allows the introduction of a geometric phase for the quantum state. Let λ be the overall phase of the wave function; we took this parameter to be $\arg(\psi_N)$ in the above discussion. We decouple this overall phase by writing $\psi_j = e^{i\lambda}\phi_j$; then, ϕ_j belongs to the so-called projective Hilbert space. From (2.25), we obtain

$$\begin{aligned} i\frac{d\psi_j}{dt} &= i\frac{de^{i\lambda}}{dt}\phi_j + i\frac{d\phi_j}{dt}e^{i\lambda} = -\frac{d\lambda}{dt}e^{i\lambda}\phi_j + i\frac{d\phi_j}{dt}e^{i\lambda} \\ &= \frac{\partial \mathscr{H}}{\partial \psi_j^*} = \frac{\partial \mathscr{H}}{\partial \phi_j}\frac{\partial \phi_j}{\partial \psi_j^*} + \frac{\partial \mathscr{H}}{\partial \phi_j^*}\frac{\partial \phi_j^*}{\partial \psi_j^*} = \frac{\partial \mathscr{H}}{\partial \phi_j^*}e^{i\lambda}; \end{aligned} \tag{2.36}$$

that is,

$$\frac{d\lambda}{dt}\phi_j = i\frac{d\phi_j}{dt} - \frac{\partial \mathscr{H}}{\partial \phi_j^*}. \tag{2.37}$$

Multiplying the above equation by ϕ_j^* and summing over j from 1 to N, we obtain

$$\frac{d\lambda(t)}{dt} = \sum_{j=1}^{N} \phi_j^* i \frac{\partial}{\partial t}\phi_j - \sum_{j=1}^{N} \phi_j^* \frac{\partial}{\partial \phi_j^*}\mathscr{H}. \tag{2.38}$$

For a linear quantum system, the second term is the same as the energy; the corresponding time integral gives the so-called dynamical phase. The time integral of the first term, $\int_0^\tau dt \sum_{j=1}^N \phi_j^* i \frac{\partial}{\partial t}\phi_j$, gives an additional contribution to the overall phase and is called the Aharonov-Anandan phase. We retain this term for the nonlinear case [14], and we show that this geometric phase plays the role of the classical action in the canonical dynamics and is thus an adiabatic invariant [11].

2.2.2 Eigenstates

For an instantaneous $\boldsymbol{R}$ (quasi-static), the quantum eigenstate is defined by

$$\begin{bmatrix} H_{11} & H_{12} & \cdots & H_{1N} \\ H_{21} & H_{22} & \cdots & H_{2N} \\ \vdots & \vdots & \ddots & \vdots \\ H_{N1} & H_{N2} & \cdots & H_{NN} \end{bmatrix} \begin{pmatrix} \psi_1 \\ \psi_2 \\ \vdots \\ \psi_N \end{pmatrix} = \mu \begin{pmatrix} \psi_1 \\ \psi_2 \\ \vdots \\ \psi_N \end{pmatrix}, \tag{2.39}$$

where μ is the chemical potential [17]. In addition to the ground state, which has clear physical significance, the higher eigenstates (also called nonlinear coherent modes) can be prepared experimentally [18]. According to Eqs. (2.24) and (2.39), one obtains $i\frac{d}{dt}\psi = \mu\psi$. Thus, for an eigenstate, its time evolution involves only a phase, $\psi(t) = \mathrm{e}^{-i\mu t}\psi(0)$. With the erasure of the overall phase, the projected state $\phi(t)$ does not change with time; i.e., $\phi(t) = \phi(0)$, or equivalently, $\frac{dq_j}{dt} = 0$ and $\frac{dp_j}{dt} = 0$ $(j = 1, \ldots, N-1)$. This condition indicates, according to Eq. (2.35), that

$$\frac{\partial \mathcal{H}}{\partial p_j} = 0, \quad \frac{\partial \mathcal{H}}{\partial q_j} = 0, \quad (j = 1, \ldots, N-1). \tag{2.40}$$

Therefore, the eigenstates correspond to the extremum energies or the fixed points of the classical Hamiltonian $\mathcal{H}$ at a given $\boldsymbol{R}$. This connection may have been known to many, e.g., see [19]. The nonlinearity in our quantum problem, however, not only makes different eigenstates nonorthogonal but also can produce more eigenstates than the dimension of the Hilbert space. Moreover, the eigenstates can be either dynamically stable or unstable. To determine the stability of the quantum eigenstates, one can linearize Eq. (2.35) about the fixed point [15] with $p_j = p_j^0 + \delta_{p_j}$ and $q_j = q_j^0 + \delta_{q_j}$, where $(p_1, q_1, \ldots, p_{N-1}, q_{N-1})$ is the fixed point, $(\delta_{p_1}, \delta_{q_1}, \ldots, \delta_{p_{N-1}}, \delta_{q_{N-1}})$ is a small perturbation about the fixed point, and $j = 1, \ldots, N$. Substituting these terms into Eq. (2.35) and retaining the Taylor expansion up to the first-order term, one obtains

$$\frac{d}{dt}\begin{pmatrix} \delta_{p_1} \\ \delta_{q_1} \\ \vdots \\ \delta_{p_{N-1}} \\ \delta_{q_{N-1}} \end{pmatrix} = \begin{bmatrix} -\frac{\partial^2 \mathscr{H}}{\partial q_1 \partial p_1} & -\frac{\partial^2 \mathscr{H}}{\partial q_1^2} & \cdots & -\frac{\partial^2 \mathscr{H}}{\partial q_1 \partial p_{N-1}} & -\frac{\partial^2 \mathscr{H}}{\partial q_1 \partial q_{N-1}} \\ \frac{\partial^2 \mathscr{H}}{\partial p_1^2} & \frac{\partial^2 \mathscr{H}}{\partial p_1 \partial q_1} & \cdots & \frac{\partial^2 \mathscr{H}}{\partial p_1 \partial p_{N-1}} & \frac{\partial^2 \mathscr{H}}{\partial p_1 \partial q_{N-1}} \\ \vdots & \vdots & \ddots & \vdots & \vdots \\ -\frac{\partial^2 \mathscr{H}}{\partial q_{N-1} \partial p_1} & -\frac{\partial^2 \mathscr{H}}{\partial q_{N-1} \partial q_1} & \cdots & -\frac{\partial^2 \mathscr{H}}{\partial q_{N-1} \partial p_{N-1}} & -\frac{\partial^2 \mathscr{H}}{\partial q_{N-1}^2} \\ \frac{\partial^2 \mathscr{H}}{\partial p_{N-1} \partial p_1} & \frac{\partial^2 \mathscr{H}}{\partial p_{N-1} \partial q_1} & \cdots & \frac{\partial^2 \mathscr{H}}{\partial p_{N-1}^2} & \frac{\partial^2 \mathscr{H}}{\partial p_{N-1} \partial q_{N-1}} \end{bmatrix} \begin{pmatrix} \delta_{p_1} \\ \delta_{q_1} \\ \vdots \\ \delta_{p_{N-1}} \\ \delta_{q_{N-1}} \end{pmatrix}. \tag{2.41}$$

The above matrix is often termed the Hamiltonian-Jacobi matrix H_J. The stability of the fixed points is then determined by the property of the eigenvalue of H_J, which is identical to the Bogoliubov excitation spectrum of the corresponding eigenstate.

If all eigenvalues of H_J are purely imaginary, one calls the fixed point an elliptic point and anticipates it following adiabatically the control parameter provided that the latter changes slowly relative to the fundamental frequencies of periodic orbits around the fixed point. In linear quantum mechanics, these frequencies are just the level spacings; thus, the breakdown of adiabaticity occurs by level crossing. In the nonlinear quantum problem, the fundamental frequencies are generally different from the level spacings; thus, adiabaticity can often be maintained even if the energy levels cross [11].

If some of the eigenvalues of H_J are complex, however, then these additional eigenstates correspond to hyperbolic points in the classical dynamics, characterized by complex fundamental frequencies and high sensitivity to small perturbations. One thus expects that such eigenstates cannot adiabatically follow the control parameter. The complex fundamental frequencies correspond to a complex Bogoliubov spectrum in addition to the mean-field solution, signifying spontaneous production of quasiparticles. In practice, we need to traverse the dynamical instability regime fast enough to avoid the quasiparticle production and slow enough to avoid nonadiabatic excitations [20].

2.2.3 *Cyclic and Quasicyclic States*

Relative to eigenstates, the adiabatic evolution of noneigenstates is in general very complex, as the motions given by (2.35) may be chaotic [19]. We choose to focus on the states around an elliptic point, where the classical orbits are regular. Here, the motions are confined on the $(N-1)$-dimensional torus, and we introduce a set of action-angle variables, $\boldsymbol{I} = (I_1, I_2, \ldots, I_{N-1})$ and $\boldsymbol{\Theta} = (\Theta_1, \Theta_2, \ldots, \Theta_{N-1})$ [15]. The angular variables change in time with frequencies $\boldsymbol{\omega} = (\omega_1, \omega_2, \ldots, \omega_{N-1})$, while the actions $\boldsymbol{I}$ are constants. Notably, according to the classical adiabatic theorem [15], the actions $\boldsymbol{I}$ are adiabatic invariants in the sense that they remain constant even if the control parameter $\boldsymbol{R}$ changes (slowly) in time. The existence of these adiabatic invariants strongly constrains the motion and ensures that a state initially close to an eigenstate (elliptic point) remains close as the system is changed slowly.

Furthermore, we can attach a physical meaning to these adiabatic invariants in the effective classical description by making a connection to the Aharonov-Anandan (AA) phase of the states. The AA phase is defined as the time integral of the first term in (2.38) for a periodic orbit,

$$\gamma_{AA}(\boldsymbol{R}) = \int_0^{\tau} dt \sum_{j=1}^{N} \phi_j^* i \frac{\partial}{\partial t} \phi_j. \tag{2.42}$$

We can rewrite this expression with the canonical variables (q_j, p_j) and further with the action-angle variables

$$\gamma_{AA} = \sum_{j=1}^{N} \int_0^{\tau} p_j dq_j = \boldsymbol{I} \cdot \boldsymbol{\Omega}, \tag{2.43}$$

where $\boldsymbol{\Omega} = (\omega_1 \tau, \omega_2 \tau, \ldots, \omega_{N-1} \tau)$ and τ is a time period. Therefore, the actions are related to the AA phase γ_{AA}, which is an observable physical quantity [21]. In the special case of $N = 2$, there is only one independent action, so the AA phase is simply $\gamma_{AA} = 2\pi I$. This simple connection can be expanded to the general case of $N > 2$, where one can identify a particular cyclic state that involves only one action I_j. For this cyclic state, we again have the simple relation $\gamma_{AA} = 2\pi I_j$. For quasiperiodic motions, where the frequencies ω_i are not commensurate, one can use rational numbers n/m to approach the ratios ω_i/ω_j as closely as possible. In this spirit, (2.43) also holds for quasicyclic motion [11].

How do the above adiabatic invariants connect to the familiar notions in the standard linear quantum mechanics? Consider the time evolution of a general state in a linear quantum system for a given $\boldsymbol{R}$, i.e., $\psi_j(t) = \psi_j(0)e^{-iE_j t}$ with $j = 1, 2, \ldots, N$, where E_j are the eigenenergies. These are (quasi-)cyclic states with the projective wave functions given by $\Phi_j(t) = \psi_j(0)e^{-i(E_j - E_N)t}$ with $j = 1, 2, \ldots, (N-1)$ and $\Phi_N(t) = \psi_N(0)$. Their AA phases can be computed from (2.42); through comparison with (2.43), we immediately find that $I_j = |\psi_j(0)|^2$. Therefore, in linear quantum mechanics, these adiabatically invariant actions I_j represent the probabilities on the energy levels. In this way, we have again derived the adiabatic theorem of linear quantum mechanics. This expression is different from the semiclassical relation discussed in [22].

2.2.4 *Two-Level Model Illustration*

As an illustration, let us consider a nonlinear two-level model,

$$\begin{aligned} i\frac{\partial \psi_1}{\partial t} &= \left[\frac{R}{2} + \frac{c}{2}\left(|\psi_2|^2 - |\psi_1|^2\right)\right]\psi_1 + \frac{v}{2}\psi_2, \\ i\frac{\partial \psi_2}{\partial t} &= \frac{v}{2}\psi_1 - \left[\frac{R}{2} + \frac{c}{2}\left(|\psi_2|^2 - |\psi_1|^2\right)\right]\psi_2. \end{aligned} \tag{2.44}$$

This model was proposed to describe the tunneling of a BEC in an optical lattice [23] or in a double-well potential [24]. The parameter c characterizes the interaction strength between atoms; v is the coupling strength between the two modes. The parameter R can be the Bloch wave number or energy difference between the two wells. We are interested in the tunneling between the energy levels shown in the top panels of Fig. 2.2 when R is increased slowly from the far negative end to the far positive end [11].

Following our general formalism, we choose the total phase as $\lambda = \arg(\psi_2)$ and introduce a pair of canonical variables, $q = \arg(\psi_1) - \arg(\psi_2)$ and $p = |\psi_1|^2$. The total energy of the system is $\mathscr{H} = \frac{v}{2}(\psi_1^*\psi_2 + \psi_2^*\psi_1) + \frac{R}{2}(|\psi_1|^2 - |\psi_2|^2) - \frac{c}{4}(|\psi_1|^2 - |\psi_2|^2)^2$. The wave function without the total phase takes the form $\phi_1 = \sqrt{p}e^{iq}$ and $\phi_2 = \sqrt{1-p}$. We then have the equivalent classical Hamiltonian as in (2.35),

$$\mathscr{H} = v\sqrt{p(1-p)}\cos q + \frac{R}{2}(2p-1) - \frac{c}{4}(2p-1)^2. \tag{2.45}$$

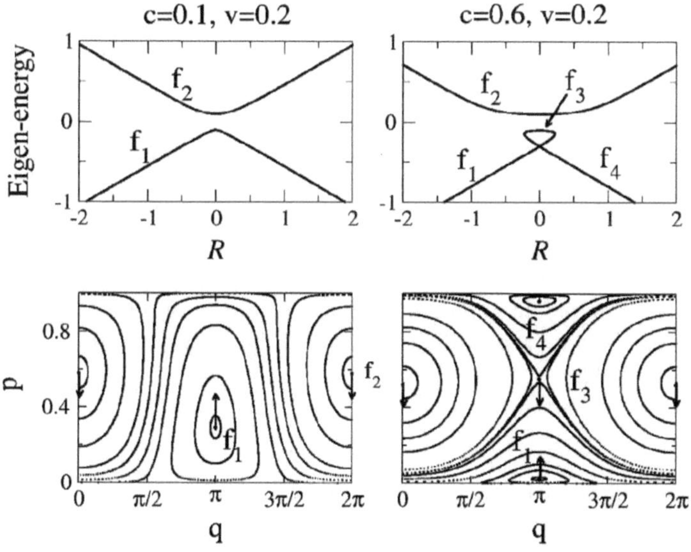

Fig. 2.2 The two top panels show the eigenenergies as a function of R for the two cases $c < v$ and $c > v$. The two bottom panels show the corresponding phase space portraits at a given value of $R = 0.05$. The arrows on the fixed points indicate the directions of their movements as R increases.

In Fig. 2.2, the structure of the eigenenergy levels of (2.44) is shown in the top panels, while the phase space orbits of the corresponding classical system (2.45) are shown in the bottom panels. When $c < v$, there are only two eigenstates and two fixed points. Since both f_1 and f_2 are elliptic with finite fundamental frequencies, the corresponding quantum states are expected to follow adiabatically with R. This finding is corroborated by our numerical simulations [11].

When $c > v$, there are two additional eigenstates, which form a loop structure; in the phase space portrait, two more fixed points appear, with one of them, f_3, being hyperbolic. Because of this structural change, the adiabatic evolution becomes very different here. First, the eigenstate corresponding to f_3 cannot follow the adiabatic change of R since f_3 is hyperbolic. This result is confirmed by our numerical integration of (2.44). Second, the fixed point f_1 can annihilate itself by colliding with f_3 as R changes slowly, leading to the breakdown of adiabaticity of the tunneling as reported numerically in [23]. Finally, we note that there is a level crossing between f_1 and f_4 at $R = 0$; however, our calculation shows that their fundamental frequencies are $v[(c/v)^2 - 1]^{1/2} \neq 0$. This result clearly illustrates our statement in the general formalism that the fundamental frequencies are not related to the level spacing in the nonlinear case.

In the abovementioned two-level model, at the two ends with $|R| \gg c$, the nonlinear term can be ignored, and the system is effectively linear. For $c < v$, where all fixed points are elliptic, the fundamental frequency ω for the periodic orbit remains finite, and the AA phase (action) is conserved [see lines (a) in the right panels of Fig. 2.3]. The initial and final probabilities on each level are indeed the same (see Fig. 2.3a), although they oscillate in the intermediate range of the parameter where the system is nonlinear. As the nonlinearity becomes substantial, the occurrence of tunneling begins to depend on the choice of the initial state. In Fig. 2.3b, where one starts with probability $I = 0.1$ on level two, tunneling occurs; however, in Fig. 2.3c, where one starts with probability $I = 0.8$ on level two, no tunneling occurs. The difference is whether a collision occurs with the hyperbolic point f_3. In Fig. 2.3b, the initial noneigenstate falls on a periodic orbit surrounding the fixed point f_1, which later collides with the hyperbolic point f_3, where the fundamental frequency drops to zero and the AA phase exhibits a finite jump [see lines (b) in the right panels of Fig. 2.3]. The jump height is proportional to the tunneling probability. In Fig. 2.3c, the initial state falls on a periodic orbit around the fixed point f_2, which does not collide with f_3 [11].

2.3 Nonlinear Adiabatic Geometric Phase

During the past decades, the Berry phase and related geometric phases [14, 25] have received renewed interest due to their important use in the implementation of quantum computing gates [26] and applications in condensed-matter physics [27]. For a nonlinear quantum system described by the nonlinear Schrödinger equations, however, the Berry phase issue is still far from being thoroughly understood. Historically,

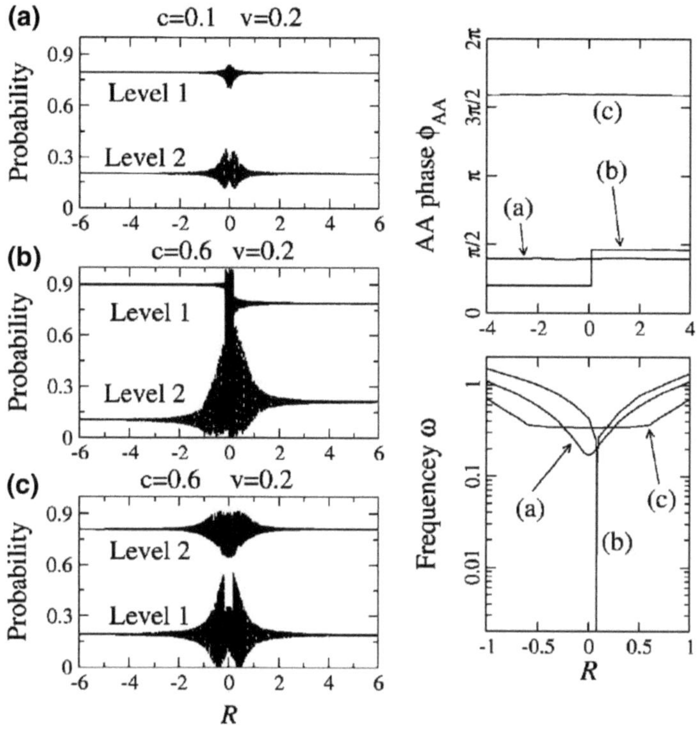

Fig. 2.3 The left panels show the change of probabilities on the two levels with R, which changes with the rate $\alpha = 0.0001$, for three different cases. The right panels show how the AA phases and the fundamental frequencies change with R in these three cases.

the geometric phase for the cyclic evolution with a finite time duration T (i.e., nonadiabatic motion) in nonlinear systems was studied many years ago [28]. It was found that the nonadiabatic geometric phase takes the form of $-\int_0^T i\langle\phi|\frac{\partial}{\partial t}|\phi\rangle dt$, analogous to its linear counterpart [14]. Here, ϕ is the wave function in the projective Hilbert space satisfying the cyclic requirement $\phi(t=0)=\phi(t=T)$. A similar formula was deduced in investigating the geometric phase in a Bose-Einstein-Josephson junction [29]. However, the Berry phase associated with the adiabatic evolution of an eigenstate in nonlinear systems is still uncharacterized. In line with the considerations of [28], one might imagine that when the parameter $\boldsymbol{R}$ vector moves in a circuit adiabatically, the adiabatic geometric phase acquired by an eigenstate takes the usual form of $-\oint i\langle\phi(\boldsymbol{R})|\frac{\partial}{\partial \boldsymbol{R}}|\phi(\boldsymbol{R})\rangle d\boldsymbol{R}$ [30]. Here, $\phi(\boldsymbol{R})$ denotes one eigenstate of the nonlinear system, and $i\langle\phi(\boldsymbol{R})|\frac{\partial}{\partial \boldsymbol{R}}|\phi(\boldsymbol{R})\rangle$ is called the Berry connection. The expression indicates that the Berry phase still equals the circuit integral of the Berry connection. In a study of the Berry phase for a specific BEC system described by a nonlocal Gross-Pitaevskii (nonlinear) equation with a quadratic potential, the exact solutions were constructed, and the Berry phase was calculated in explicit form [31].

The obtained adiabatic geometric phase, however, does not equal the above expression; an additional term emerges that is directly imparted by the presence of nonlinearity. The reason has not yet been determined [31], and this controversy is not properly resolved, prompting further investigation. The discrepancy may indicate that some subtle and important aspects were missed in previous theoretical considerations.

In the present section, we have made a thorough analysis of the Berry phase issue for nonlinear systems. Our analytic deduction clearly indicates that the Berry phase is dramatically modified by the nonlinearity. The underlying mechanism has been revealed: for a nonlinear system, because the Hamiltonian is a functional of the instantaneous wave functions, the Bogoliubov excitations induced by the slow change of the system are allowed to feed back to the Hamiltonian. These excitations are accumulated during an adiabatic evolution and ultimately contribute a finite phase of geometric nature [32].

2.3.1 Adiabatic Parameter Expansion

Without loss of generality, let us consider the Schrödinger equation with a quadratic nonlinear term [32],

$$i\frac{\partial\psi}{\partial t} = H_0\psi + g|\psi|^2\psi, \tag{2.46}$$

where $H_0 = -\frac{1}{2}\nabla^2 + V(\boldsymbol{R}; r)$, with $\boldsymbol{R}$ being the parameter vector that varies slowly in time, and g is the nonlinear parameter representing the interaction between the coherent atoms. The total energy of the system $E_T = \int dr E(\psi^*, \psi)$, where the energy density $E(\psi^*, \psi) = \psi^* H_0\psi + \frac{1}{2}g|\psi|^4$. The above system is invariant under gauge transformations of the first kind, i.e., $\psi(r, t) \to \mathrm{e}^{i\eta}\psi(r, t)$ with constant η. The gauge symmetry implies that the total atom number is conserved, i.e., $\int dr|\psi|^2 = 1$.

Let λ be the overall phase of the wave function. We may take this parameter to be the phase of the wave function at a fixed position r_0, for example, $\lambda = -\arg[\psi(r_0, t)]$. We decouple this overall phase by writing $\psi = \mathrm{e}^{-i\lambda}\phi$, where ϕ belongs to the so-called projective Hilbert space. From (2.46), we obtain

$$\frac{d\lambda}{dt} = -i\langle\phi|\frac{\partial}{\partial t}|\phi\rangle + \int dr E(\phi^*, \phi) + \frac{g}{2}\langle\phi|\phi^*\phi|\phi\rangle. \tag{2.47}$$

The eigenequation of the system is

$$H_0\bar{\psi} + g|\bar{\psi}|^2\bar{\psi} = \mu\bar{\psi}, \tag{2.48}$$

where $\bar{\psi}$ is the eigenfunction and μ is the eigenvalue (or chemical potential).

We now assume that the parameter vector $\boldsymbol{R}$ varies slowly in time, and we introduce the dimensionless adiabatic parameter of $\varepsilon \sim |d\boldsymbol{R}/dt| \sim 1/T$ as the measure of how slow the parameters change. The adiabatic parameter tends to zero, i.e., $\varepsilon \to 0$, indicating the adiabatic limit. T is the time duration.

Consequently, the expression of the total phase can be expanded in a perturbation series in the adiabatic parameter, i.e.,

$$\frac{d\lambda}{dt} = \alpha_0(\varepsilon^0) + \alpha_1(\varepsilon^1) + O(\varepsilon^2). \tag{2.49}$$

When the parameters move in a circuit, the eigenstate evolves for an infinitely long time duration in the adiabatic limit. The time integral of the zero-order term gives the so-called dynamical phase because it is closely related to the temporal process of the evolution. The time integral of the first-order term makes an additional contribution to the overall phase, which will later be shown to be of a geometric nature; that is, it depends only on the geometry of the closed path in the parameter space. The contribution of the higher-order term vanishes in the adiabatic limit.

In the quantum evolution with slowly changing parameters, we assume that $\phi = \bar{\phi}(\boldsymbol{R}) + \delta\phi(\boldsymbol{R})$, where $\bar{\phi}(\boldsymbol{R})$ is the wave function of the instantaneous eigenstate corresponding to the local minimum energy. $\delta\phi(\boldsymbol{R})$ denotes the secular part of the Bogoliubov excitations induced by the system's slow change, while the rapid oscillations in the excitations are ignored because they vanish after long-term averaging. $\delta\phi(\boldsymbol{R})$ depends on the adiabatic parameter and is of order ε; thus, from (2.47) and in conjunction with relation (2.48), we obtain the explicit expressions as follows [32]:

$$\alpha_0(\varepsilon^0) = \mu(\boldsymbol{R}), \tag{2.50}$$

$$\alpha_1(\varepsilon^1) = -i\int dr\left(\bar{\phi}^*\frac{\partial}{\partial t}\bar{\phi}\right) + g\int dr\left(\bar{\phi}^2\bar{\phi}^*\delta\phi^* + (\bar{\phi}^*)^2\bar{\phi}\delta\phi\right). \tag{2.51}$$

2.3.2 *Projective Hilbert Space Description*

From the above expressions, we observe that the dynamical phase has been modified to be the time integral of the chemical potential rather than the energy. This condition arises because the instantaneous eigenstates are fed back to the Hamiltonian. Notably, the first-order term, i.e., the Berry phase term, has been modified due to the feedback of the Bogoliubov excitations to the Hamiltonian. To evaluate the Berry phase term qualitatively and express the modified geometric phase explicitly, we introduce a set of orthogonal bases $|k\rangle$ with $k = 1, 2, \ldots, N$ and the variable ψ_j, which is the jth component, i.e., $\psi_j = \langle j|\psi\rangle$. Without loss of generality, the projective Hilbert space is set to be of a specific gauge such that the phase of the Nth component is zero. In the projective Hilbert space, the new variables (n_j, θ_j) are introduced through $\phi_j = \sqrt{n_j}e^{i\theta_j}$. Substituting the expression of $\psi_j = \sqrt{n_j}e^{i\theta_j}e^{-i\int^t \beta dt}$ into the nonlinear equation and separating the real and imaginary parts, we obtain the following differential equations for the density n_j and the phase θ_j [32]:

$$\frac{dn_j}{dt} = f_j, \quad \frac{d\theta_j}{dt} = h_j, \quad j = 1, 2, \ldots, N-1, \tag{2.52}$$

where f_j and h_j are functions of the amplitudes, relative phases, matrix elements $C_{jk}(\boldsymbol{R}) = \langle j|H_0(\boldsymbol{R})|k\rangle$, and the overlap integral $D_{j,k,l,m} = \langle j|\langle k|l\rangle|m\rangle$. Their explicit expressions can be readily deduced but are not given here.

The norm conservation condition $n_N = 1 - \sum_{k=1}^{N-1} n_k$ has been used to remove the variable n_N in the above equations. In the representation of new variables, $(\bar{n}_j, \bar{\theta}_j)$ satisfy equations of the equilibrium state, i.e.,

$$\left(\frac{dn_j}{dt}, \frac{d\theta_j}{dt}\right)\bigg|_{(\bar{n}_j, \bar{\theta}_j)} = 0, \tag{2.53}$$

where $(\bar{n}_j, \bar{\theta}_j)$ are functions of the parameter $\boldsymbol{R}$ corresponding to the eigenstates of the system. We make a perturbation expansion around the eigenstate with

$$n_j = \bar{n}_j(\boldsymbol{R}) + \delta n_j(\boldsymbol{R}), \quad \theta_j = \bar{\theta}_j(\boldsymbol{R}) + \delta\theta_j(\boldsymbol{R}). \tag{2.54}$$

Here, $\bar{\phi}_j(\boldsymbol{R}) = \sqrt{\bar{n}_j(\boldsymbol{R})}\mathrm{e}^{i\bar{\theta}_j(\boldsymbol{R})}$, $\delta n_j(\boldsymbol{R})$ and $\delta\theta_j(\boldsymbol{R})$ are the excitations of order ε. Then, inserting the above expansion into equations (2.52), ignoring higher-order terms such as $\partial\delta n_j/\partial t$ and $\partial\delta\theta_j/\partial t$, and denoting $v = (n_1, \theta_1; \cdots; n_{N-1}, \theta_{N-1})$, we obtain

$$\frac{d\bar{v}}{d\boldsymbol{R}} \cdot \frac{d\boldsymbol{R}}{dt} = \mathscr{L}\delta v, \tag{2.55}$$

where the matrix takes the form

$$\mathscr{L} = \{\mathscr{L}_{jk}\}_{(N-1,N-1)}, \quad \mathscr{L}_{jk} = \begin{pmatrix} \frac{\partial f_j}{\partial n_k} & \frac{\partial f_j}{\partial \theta_k} \\ \frac{\partial h_j}{\partial n_k} & \frac{\partial h_j}{\partial \theta_k} \end{pmatrix}_{v=\bar{v}}. \tag{2.56}$$

Then, inversely, we have

$$\delta v = \mathscr{L}^{-1} \cdot \frac{d\bar{v}}{d\boldsymbol{R}} \cdot \frac{d\boldsymbol{R}}{dt}. \tag{2.57}$$

The differential relation between the new variables and old ones takes the form

$$\begin{pmatrix} \delta\phi_j \\ \delta\phi_j^* \end{pmatrix} = \Pi_j \begin{pmatrix} \delta n_j \\ \delta\theta_j \end{pmatrix}, \tag{2.58}$$

in which

$$\Pi_j = \begin{pmatrix} \frac{1}{2\sqrt{\bar{n}_j}}\mathrm{e}^{i\bar{\theta}_j} & i\sqrt{\bar{n}_j}\mathrm{e}^{i\bar{\theta}_j} \\ \frac{1}{2\sqrt{\bar{n}_j}}\mathrm{e}^{-i\bar{\theta}_j} & -i\sqrt{\bar{n}_j}\mathrm{e}^{-i\bar{\theta}_j} \end{pmatrix}. \tag{2.59}$$

2.3.3 Nonlinear Adiabatic Geometric Phase

Substituting (2.57) and (2.58) into (2.51), we obtain the explicit expression of the adiabatic geometric phase that contains two terms [32],

$$\gamma_g = \gamma_B + \gamma_{NL}, \tag{2.60}$$

where the first term is the usual Berry phase formula, i.e.,

$$\gamma_B = -i \oint \langle \bar{\phi} | \nabla_{\boldsymbol{R}} | \bar{\phi} \rangle \cdot d\boldsymbol{R} = \oint \sum_{j=1}^{N-1} \bar{n}_j \cdot \frac{\partial \bar{\theta}_j}{\partial \boldsymbol{R}} d\boldsymbol{R}, \tag{2.61}$$

and the additional term is from the nonlinearity, taking the form

$$\gamma_{NL} = g \oint \langle \Lambda | \Pi \cdot \mathscr{L}^{-1} | \frac{d\bar{v}}{d\boldsymbol{R}} \rangle \cdot d\boldsymbol{R}. \tag{2.62}$$

Here,

$$\Lambda = \left((\bar{n}_1 + \sum_{j=1}^{N-1} \bar{n}_j - 1)\sqrt{\bar{n}_1} \mathrm{e}^{i\bar{\theta}_1}, (\bar{n}_1 + \sum_{j=1}^{N-1} \bar{n}_j - 1)\sqrt{\bar{n}_1} \mathrm{e}^{-i\bar{\theta}_1}, \ldots \right),$$

$$\frac{d\bar{v}}{d\boldsymbol{R}} = \left(\frac{d\bar{n}_1}{d\boldsymbol{R}}, \frac{d\bar{\theta}_1}{d\boldsymbol{R}}, \ldots, \frac{d\bar{n}_{N-1}}{d\boldsymbol{R}}, \frac{d\bar{\theta}_{N-1}}{d\boldsymbol{R}} \right)^T,$$

and diagonal matrix $\Pi = \mathrm{diag}(\Pi_1, \Pi_2, \ldots, \Pi_{N-1})$. Note that to simplify the expression of Λ, we use the approximation that the overlap integral $D_{j,k,l,m} \simeq 0$ when the subscripts are not all identical.

Both γ_B and γ_{NL} have the geometric property of the parameter space. The novel second term indicates that the Bogoliubov excitations induced by the slow change of the system, which is negligible in the linear case, can be accumulated in the nonlinear adiabatic evolution and can contribute to the finite phase of a geometric nature [32].

2.3.4 Two-Mode Model Illustration

As an illustration of our theoretical formalism, we consider the simple two-mode BEC model described by the nonlinear equation [33]

$$i\frac{d}{dt}\begin{pmatrix} \Psi_1 \\ \Psi_2 \end{pmatrix} = H(\Psi_1, \Psi_2)\begin{pmatrix} \Psi_1 \\ \Psi_2 \end{pmatrix}, \tag{2.63}$$

with

$$H(\Psi_1, \Psi_2) = \begin{pmatrix} Z|\Psi_1|^2 & \frac{\rho}{2}e^{-i\varphi} \\ \frac{\rho}{2}e^{i\varphi} & Z|\Psi_2|^2 \end{pmatrix}, \tag{2.64}$$

and $\boldsymbol{R} = (X = \rho\cos\varphi, Y = \rho\sin\varphi, Z)$ are parameters. For simplicity, we fix ρ and Z and change the parameter φ from 0 to 2π adiabatically.

The eigenequations read

$$H(\bar{\Phi}_1, \bar{\Phi}_2)\begin{pmatrix} \bar{\Phi}_1 \\ \bar{\Phi}_2 \end{pmatrix} = \mu\begin{pmatrix} \bar{\Phi}_1 \\ \bar{\Phi}_2 \end{pmatrix}. \tag{2.65}$$

For the nonlinear system, the number of eigenstates may be larger than the dimension of the Hilbert space, and the eigenstates may be unstable [11]. We have obtained four eigenstates for the case $Z > \rho$ by solving the above eigenequations. Three of these eigenstates are stable, and one is unstable. We choose the following stable eigenstate to illustrate our theory:

$$\begin{aligned} \bar{\Phi}_1 &= \sqrt{\frac{1}{2}\left(1 - \sqrt{1 - \frac{\rho^2}{Z^2}}\right)}, \\ \bar{\Phi}_2 &= e^{i\varphi}\sqrt{\frac{1}{2}\left(1 + \sqrt{1 - \frac{\rho^2}{Z^2}}\right)}, \end{aligned} \tag{2.66}$$

with the eigenvalue of $\mu = Z$.

The Berry's term of the geometric phase is readily deduced:

$$\gamma_B = -i\int_0^{2\pi}\left(\bar{\Phi}_2^*\frac{\partial}{\partial\varphi}\bar{\Phi}_2\right)d\varphi = \pi\left(1 + \sqrt{1 - \frac{\rho^2}{Z^2}}\right). \tag{2.67}$$

We now derive the additional term γ_{NL} of the geometric phase. Let us introduce the new variables (n, θ) through $(\Phi_1, \Phi_2) = (\sqrt{1-n}, \sqrt{n}e^{i\theta})$. Substituting $(\Psi_1, \Psi_2) = e^{-i\int^t \beta dt(\Phi_1, \Phi_2)}$ into (2.63) and separating the real and the imaginary parts, we obtain four differential equations, two of which are identical due to the norm conservation:

$$\frac{dn}{dt} = -\rho\sqrt{n - n^2}\sin(\theta - \varphi), \tag{2.68}$$

$$\frac{d\theta}{dt} = -\frac{\rho\sqrt{1-n}}{2\sqrt{n}}\cos(\theta - \varphi) - Zn + \beta, \tag{2.69}$$

$$\beta = Z(1-n) + \frac{\rho}{2}\sqrt{\frac{n}{1-n}}\cos(\theta - \varphi). \tag{2.70}$$

The eigenstate comprises the fixed point of (2.68) and (2.69), i.e.,

$$\bar{n} = \frac{1}{2}\left(1 + \sqrt{1 - \frac{\rho^2}{Z^2}}\right), \quad \bar{\theta} = \varphi. \tag{2.71}$$

We perform a perturbation expansion around the eigenstate with $n = \bar{n}(\varphi) + \delta n$ and $\theta = \bar{\theta}(\varphi) + \delta\theta$. Then, inserting the above expansion into (2.68) and (2.69), we obtain

$$\begin{pmatrix} \frac{\partial \bar{n}}{\partial \varphi} \\ \frac{\partial \bar{\theta}}{\partial \varphi} \end{pmatrix} \frac{d\varphi}{dt} = \mathscr{L} \begin{pmatrix} \delta n \\ \delta \theta \end{pmatrix}, \tag{2.72}$$

with

$$\mathscr{L} = \begin{pmatrix} 0 & -\rho\sqrt{\bar{n} - \bar{n}^2} \\ -2Z + \frac{\rho}{4(\bar{n}-\bar{n}^2)^{3/2}} & 0 \end{pmatrix}. \tag{2.73}$$

Then, we have

$$\gamma_{NL} = Z \int_0^{2\pi} \langle \Lambda | \Pi \cdot \mathscr{L}^{-1} \cdot (\frac{\partial \bar{n}}{\partial \varphi}, \frac{\partial \bar{\theta}}{\partial \varphi})^T \rangle \cdot d\varphi, \tag{2.74}$$

where $\Lambda = ((2\bar{n} - 1)\sqrt{\bar{n}}\mathrm{e}^{-i\theta}, (2\bar{n} - 1)\sqrt{\bar{n}}\mathrm{e}^{i\theta})$. Finally, we obtain

$$\gamma_{NL} = \frac{\pi \rho^2}{Z\sqrt{Z^2 - \rho^2}}. \tag{2.75}$$

Combining formulas (2.67) and (2.74), we arrive at an explicit expression of the adiabatic geometric phase:

$$\gamma_g = \pi \left(1 + \sqrt{1 - \frac{\rho^2}{Z^2}} + \frac{\rho^2}{Z\sqrt{Z^2 - \rho^2}}\right). \tag{2.76}$$

This new geometric phase has been verified by numerically solving the nonlinear Schödinger equation (2.63). This phase can also be interpreted as the flux of a virtual magnetic field $\boldsymbol{M}$ through the surface enclosed by the close circuit in the parameter space. The virtual field has been deduced to take the form of [32]

$$\boldsymbol{M} = \frac{\boldsymbol{R}}{2(Z^2 - \rho^2)^{3/2}}, \tag{2.77}$$

in contrast to the Dirac monopole field $\boldsymbol{R}/(2R^3)$ described by Berry for the linear case [34].

We now discuss the geometric meaning of the new phase. The state vector ψ can be parametrized in a Bloch sphere according to

$$|\psi\rangle\langle\psi| = \frac{1}{2}(\boldsymbol{I} + \boldsymbol{s}\cdot\boldsymbol{\sigma}), \tag{2.78}$$

where $|\psi\rangle = (\cos\frac{\alpha}{2}\mathrm{e}^{-i\delta/2}, \sin\frac{\alpha}{2}\mathrm{e}^{i\delta/2})$, $\boldsymbol{s} = (\sin\alpha\cos\delta, \sin\alpha\sin\delta, \cos\alpha)$, and $\boldsymbol{I}$ and $\boldsymbol{\sigma}$ are the unit and Pauli matrices, respectively. In the new variables, the nonlinear phase can be expressed as a function of the solid angle in a Bloch sphere [i.e., $\Omega_B = 2\pi(1-\cos\alpha)$] and the solid angle in the parameter space [i.e., $\Omega_P = 2\pi(1 - Z/\sqrt{\rho^2+Z^2})$]. The two angles are typically not identical in the nonlinear case. To understand this condition, we first consider the linear case $Z = 0$. The system can be viewed as a spin $\boldsymbol{s}$ driven by an external magnetic field $\boldsymbol{B} = (\rho\cos\varphi, \rho\sin\varphi, Z)$. When the magnetic field varies in time adiabatically, the dynamical equation

$$\frac{d\boldsymbol{s}}{dt} = \boldsymbol{B}\times\boldsymbol{s} \tag{2.79}$$

indicates that the spin remains parallel to the field. In this case, $\Omega_B = \Omega_P$. In the presence of the nonlinearity, the magnetic field is modulated self-consistently by the spin, and the effective field $\boldsymbol{B}^* = (\rho\cos\varphi, \rho\sin\varphi, Z\cos\alpha)$. This result is presented in (2.64), where parameter Z has been renormalized by $|\Psi_1|^2 - |\Psi_2|^2 = \cos\alpha$. The adiabatic evolution of the spin is expected to remain parallel to the modulated field $\boldsymbol{B}^*$ rather than $\boldsymbol{B}$. Thus, for a cyclic adiabatic evolution of the spin, the solid angles in the Bloch sphere and the parameter space are generally not identical. On the other hand, from (2.67) and (2.74) and using the relations $\bar{n} = \Omega_B/(4\pi)$ and $\rho/Z = \tan[\arccos(1-\Omega_P/2\pi)]$, we can express our nonlinear geometric phase in terms of these solid angles, i.e.,

$$\gamma_g = \frac{\Omega_B}{2} + \frac{\pi(1-\frac{\Omega_B}{2\pi})(1-\frac{\Omega_P}{2\pi})}{(1-\frac{\Omega_P}{2\pi}) - \sqrt{(2-\frac{\Omega_P}{2\pi})(\frac{\Omega_P}{2\pi})/[(2-\frac{\Omega_B}{2\pi})(\frac{\Omega_B}{2\pi})]^{3/2}}}. \tag{2.80}$$

For the linear case $Z = 0$, $\Omega_P = 2\pi$, the above expression reduces to the well-known relation $\gamma_g = \Omega_B/2$, i.e., the Berry phase equals half the solid angle.

Finally, we discuss these results. Let us first recall the nonadiabatic geometric phase for a cyclic motion. By inserting $\psi = \mathrm{e}^{-i\lambda}\phi$ into the Schrödinger equation $i\frac{\partial\psi}{\partial t} = H\psi$, one can readily obtain that $\frac{d\lambda}{dt} = -i\langle\phi|\frac{\partial}{\partial t}|\phi\rangle + \langle\phi|H|\phi\rangle$. Here, ϕ is the wave function in the projective Hilbert space satisfying the cyclic requirement $\phi(t=0) = \phi(t=T)$. The total phase acquired during the cyclic evolution contains two parts: $-\int_0^T i\langle\phi|\frac{\partial}{\partial t}|\phi\rangle dt$ and $\int_0^T \langle\phi|H|\phi\rangle dt$. The former has no relation to Hamiltonian H and can therefore be regarded as the geometric phase for the cyclic state [28]. The above deduction is correct regardless of nonlinearity, requiring only that the system be invariant under a gauge transformation (the $H = H_0 + g|\psi|^2$ discussed in this work has this symmetry, for example). A question arises as to why the above simple deduction does not apply to our adiabatic case. The main reason is that the adiabatic limit is a process, i.e., one can approach this limit but cannot reach it. That is, for any small adiabatic parameter ε, i.e., $\boldsymbol{R}$ sweeping at any small rate, the dynamical solution obtained by solving the Schrödinger equation deviates from the adiabatic solution by a small quantity of order ε. This point has been clarified by the discussion in previous sections and expressed explicitly by $\phi = \bar{\phi}(\boldsymbol{R}) + \delta\phi(\boldsymbol{R})$. Moreover, as revealed by our previous discussion, in the presence of nonlinearity, this infinitesimal deviation can accumulate to contribute a finite phase during an infinitely long time evolution $T \sim 1/\varepsilon \to \infty$. In contrast, for the cyclic evolution with a finite time duration, this kind of accumulation cannot emerge. Therefore, the adiabaticity is crucial to the emergence of our nonlinear correction [32].

References

1. Weinberg, S.: Phys. Rev. Lett. **62**, 485 (1989)
2. Pethick, C.J., Smith, H.: Bose-Einstein Condensation in Dilute Gases. Cambridge University Press, London (2002)
3. Gross, E., Nuovo Cimento, P.: **20**, 454 (1961); Pitaevskii, L.P.: Zh. Eksp. Teor. Fiz. **40**, 646 (1961); Pitaevskii, L.P.: Sov. Phys. JETP **13**, 451 (1961)
4. Chen, Y., Haus, H.A.: Chaos **10**, 529 (2000)
5. Wai, P.K.A., Menyuk, C.R.H., Chen, H.: Opt. Lett. **16**, 1231 (1991); Wai, P.K.A., Menyuk, C.R.: Lightwave Technol. J. **14**, 362 (1996)
6. Pu, H., Maenner, P., Zhang, W.P., Ling, H.Y.: Phys. Rev. Lett. **98**, 050406 (2007); Meng, S.Y., Fu, L.B., Liu, J.: Phys. Rev. A **78**, 053410 (2008)
7. Itin, A.P., Watanabe, S.: Phys. Rev. Lett. **99**, 223903 (2007); Itin, A.P., Watanabe, S., Konotop, V.V.: Phys. Rev. A **77**, 043610 (2008)
8. Bergmann, K., Theuer, H., Shore, B.W.: Rev. Mod. Phys. **70**, 1003 (1998)
9. Dahan, M.B.: Phys. Rev. Lett. **76**, 4508 (1996); Wilkinson, S.R., et al.: Phys. Rev. Lett. **76**, 4512 (1996)
10. Landau, L.D., Lifshitz, E.M.: Quantum Mechanics. Pergamon Press, New York (1977); Shapere, A., Wilczek, F.: Geometric Phases in Physics. World Scientific, Singapore (1989); Bohm, A., Mostafazadeh, A., Koizumi, H., Niu, Q., Zwanziger, J.: The Geometric Phase in Quantum Systems. Springer, New York (2003)
11. Liu, J., Wu, B., Niu, Q.: Phys. Rev. Lett. **90**, 170404 (2003)
12. Kivshar, Y.S., Malomed, B.A.: Rev. Mod. Phys. **61**, 763 (1989)
13. Band, Y.B., Malomed, B., Trippenbach, M.: Phys. Rev. A **65**, 033607 (2002); Band, Y.B., Trippenbach, M.: Phys. Rev. A **65**, 053602 (2002); deValcárcel, G.J. Phys. Rev. A **67**, 051601(R) (2003)
14. Aharonov, Y., Anandan, J.: Phys. Rev. Lett. **58**, 1593 (1987)
15. Arnol'd, V.I.: Mathematical Methods of Classical Mechanics. Springer, Berlin (1978); Lichtenberg, A.J., Lieberman, M.A.: Regular and Stochastic Motion. Springer, Berlin (1983)
16. Franzosi, R., Penna, V.: Phys. Rev. A **63**, 043609 (2001)
17. For a linear system, the chemical potential is equal to the eigenenergy. However, for a nonlinear system, they do not equal to each other in general
18. Yukalov, V.I., Yukalova, E.P., Bagnato, V.S.: Phys. Rev. A **66**, 043601 (2002)
19. Coullet, P., Vandenberghe, N.: J. Phys. B **35**, 1593 (2002); Franzosi, R., Penna, V.: Phys. Rev. A **65**, 013601 (2002)
20. Yurovsky, V.A., Ben-Reuven, A.: Phys. Rev. A **63**, 043404 (2001); Vardi, A., Yurovsky, V.A., Anglin, J.R.: ibid. **64**, 063611 (2001); Wu, B., Niu, Q.: ibid. **64**, 061603 (2001); Staliunas, K., Longhi, S., Anglin, J.R.: Phys. Rev. Lett. **89**, 210406 (2002)

21. Suter, D., Mueller, K.T., Pines, A.: Phys. Rev. Lett. **60**, 1218 (1988); Weinfurter, H.: ibid. **64**, 1318 (1990)
22. Berry, M.V.: J. Phys. A **17**, 1225 (1984)
23. Wu, B., Niu, Q.: Phys. Rev. A **61**, 023402 (2000); Liu, J., et al.: Phys. Rev. A **66**, 023404 (2002)
24. Milburn, G.J., et al.: Phys. Rev. A **55**, 4318 (1997); Zapata, I., Sols, F., Leggett, A.J.: ibid. **57**, R28 (1998); Vardi, A., Anglin, J.R.: ibid. **86**, 568 (2001); Kohler, S., Sols, F.: ibid. **89**, 060403 (2002)
25. Wilczek, F., Zee, A.: Phys. Rev. Lett. **52**, 2111 (1984)
26. Macchiavello, C., Palma, G.M., Zeilinger, A.: Quantum Computation and Quantum Information Theory. World Scientific, Singapore (2000); Jones, J., Vedral, V., Ekert, A.K., Castagnoli, C.: Nature (London) **403**, 869 (2000); Falci, G., Fazio, R., Palma, G.M., Siewert, J., Vedral, V.: ibid. **407**, 355 (2000); Duan, L.M., Cirac, I., Zoller, P.: Science **292**, 1695 (2001)
27. Zhang, C., Dudarev, A.M., Niu, Q.: Phys. Rev. Lett. **97**, 040401 (2006); Xiao, D., Yao, Y., Fang, Z., Niu, Q.: ibid. **97**, 026603 (2006); Yao, W., Niu, Q.: ibid. **101**, 106401 (2008)
28. Garrison, J.C., Chiao, R.Y.: Phys. Rev. Lett. **60**, 165 (1988); Anandan, J.: Phys. Rev. Lett. **60**, 2555 (1988)
29. Balakrishman, R., Metha, M.: Euro. Phys. J. D **33**, 437 (2005)
30. Wu, B., Liu, J., Niu, Q.: Phys. Rev. Lett. **94**, 140402 (2005); Yi, X.X., Huang, X.L., Wang, W.: Phys. Rev. A **77**, 052115 (2008)
31. Litvinets, F.N., et al.: J. Phys. A. **39**, 1191 (2006)
32. Liu, J., Fu, L.B.: Phys. Rev. A **81**, 052112 (2010)
33. Smerzi, A., Fantoni, S., Giovanazzi, S., Shenoy, S.R.: Phys. Rev. Lett. **79**, 4950 (1997)
34. Berry, M.V.: Proc. R. Soc. Lond. A **392**, 45 (1984)

Chapter 3
Quantum-Classical Correspondence of an Interacting Bosonic Many-Body System

Abstract In this chapter, we discuss the adiabatic limit and the semiclassical limit, and we investigate the commutability between these two limits with a second-quantized two-mode bosonic model. We deduce the adiabatic Berry phase and the classical Hannay angle of an interacting bosonic many-body system , and we discuss the relationship between the quantum Berry phase, the classical Hannay angle, and the mean-field nonlinear geometric phase of this system.

3.1 Commutability Between the Semiclassical Limit and the Adiabatic Limit

3.1.1 Hamiltonian

Let us consider a quantum system whose Hamiltonian is time dependent,

$$i\hbar\frac{\partial}{\partial t}|\psi\rangle = H(\boldsymbol{R}(\alpha t))|\psi\rangle, \tag{3.1}$$

where the adiabatic limit is the limit of slow change, i.e., $\alpha \to 0$, and the semiclassical limit refers to the limit under which the quantum theory reduces to classical mechanics, i.e., $\hbar \to 0$. As we discussed in Chaps. 1 and 2, two adiabatic theorems exist, one for classical systems and the other for quantum systems. It is then natural and interesting to ask the following: Can one connect these two adiabatic theorems under the semiclassical limit?

In 1977, Hwang and Pechukas claimed that the above two limits are equivalent [1]. Their argument is based on the scaling law that if $\psi(t, \hbar)$ is a family of solutions to Eq. (3.1) for $\alpha = 1$ and variable $\hbar$, then $\phi(t, \alpha) = \psi(\alpha t, \alpha\hbar)$ is a family of solutions to Eq. (3.1) for fixed $\hbar$ and variable α. Thus, it is expected that one can eliminate α from the above Schrödinger equation with a scaled time $\tau = \alpha t$ and an effective Planck constant $\tilde{\hbar} = \alpha\hbar$.

This point, however, was refuted by Berry [2], who noted that these two limits are not equivalent because the Hamiltonian H may depend implicitly on $\hbar$ if one recalls

J. Liu et al., *Nonlinear Adiabatic Evolution of Quantum Systems*,
https://doi.org/10.1007/978-981-13-2643-1_3

the expression of kinetic energy operator $T = \frac{\hbar^2}{2m}\nabla^2$. Moreover, Berry showed that these two limits are incommutable in a simple double-well model: the Landau-Zener (LZ) tunneling rate is zero if the adiabatic limit $\alpha \to 0$ is taken first; the rate becomes one if one takes the semiclassical limit $\hbar \to 0$ first [2].

We now revisit the commutability between the semiclassical limit and the adiabatic limit with a second-quantized two-mode tunneling model. This model can be used to describe a BEC system in which only two quantum states are important, such as in a double-well potential or a system with two internal quantum states [3, 4]. In this model, the semiclassical limit is $N \to \infty$, with N being the number of bosons. In the large N limit, the second-quantized model becomes a two-level mean-field model. We show that one can recover the second-quantized model by quantizing this mean-field model with the Sommerfeld rule. As N can be changed in experiments, the semiclassical limit becomes experimentally accessible [5].

Notably, the commutability between the two limits, $N \to \infty$ and $\alpha \to 0$, in this second-quantized model depends on its mean-field interaction strength c. If c is small, the two limits are commutable; if c exceeds a critical value, the two limits become incommutable. Such a dependence on c is related to a topological change in the structure of the energy bands.

The second-quantized two-mode model is

$$\hat{H} = \frac{\gamma}{2}(\hat{a}^\dagger \hat{a} - \hat{b}^\dagger \hat{b}) + \frac{v}{2}(\hat{a}^\dagger \hat{b} + \hat{a}\hat{b}^\dagger) - \frac{\lambda}{2}(\hat{a}^\dagger \hat{a} - \hat{b}^\dagger \hat{b})^2, \tag{3.2}$$

where generators (annihilators) $\hat{a}^\dagger$ and $\hat{b}^\dagger$ ($\hat{a}$ and $\hat{b}$) are related to two different quantum states. In the Hamiltonian $\hat{H}$, the energy offset γ between the two quantum states changes with time as $\gamma = \alpha t$. The parameter v measures the coupling between the two states, while $\lambda > 0$ is the interacting strength between bosons. The minus sign before λ indicates that the interaction is attractive. In this system, the total number of bosons N is conserved.

3.1.2 Semiclassical Limit and Adiabatic Limit

For the second-quantized model (3.2), the semiclassical limit is $N \to \infty$. In such a limit, the system's dynamic behavior is described by the following nonlinear two-level model:

$$i\frac{d}{dt}\begin{pmatrix} a \\ b \end{pmatrix} = \left\{\left[\frac{\gamma}{2} - \frac{\lambda}{2}(|a|^2 - |b|^2)\right]\sigma_z + \frac{v}{2}\sigma_x\right\}\begin{pmatrix} a \\ b \end{pmatrix}, \tag{3.3}$$

where $c = N\lambda$ and $|a|^2 + |b|^2 = 1$. This model is often called a mean-field model. To obtain the mean-field model, one focuses on the Gross-Pitaevskii (GP) states [2] $|\Psi_{GP}\rangle = \frac{1}{\sqrt{N!}}(a\hat{a}^\dagger + b\hat{b}^\dagger)^N|\text{vac}\rangle$. By computing the expectation value $\langle \hat{H} \rangle =$

$\langle \Psi_{GP}|\hat{H}|\Psi_{GP}\rangle$, one obtains the mean-field Hamiltonian $H_{mf} = \hat{H}/N$ (up to a trivial constant) in the limit of $N \to \infty$. The Hamiltonian H_{mf} leads to the dynamics in (3.3). For a rigorous account of the large N limit as a semiclassical limit in models such as (3.2), we refer readers to [6].

We emphasize that the semiclassical limit $N \to \infty$ is taken with the mean-field interaction strength $c = N\lambda$ kept constant. Physically, this procedure ensures that the series of systems with different N have the same physics. If λ were kept constant instead of c, then the last term in (3.2) would become too dominating at the large N limit, completely changing the physics of the system. When the model (3.2) is used to describe a BEC in a double-well potential, the limit $N \to \infty$ at a constant c is equivalent to having a larger trap for more atoms in the BEC or to tuning λ smaller with the Feshbach resonance technique [7].

We are interested in how the second-quantized model (3.2) behaves in the two limits $N \to \infty$ and $\alpha \to 0$ and in particular whether the model's behavior depends on which limit is taken first. For this purpose, we follow Berry's methodology [2] to focus on the tunneling behavior of the quantized model.

3.1.3 Tunneling Rates

In Fig. 3.1, the tunneling rates are plotted as a function of the mean-field interaction strength c [5]. Two sets of tunneling rates are calculated: one with the quantized model (3.2) for a fixed number of bosons and the other with the mean-field model (3.3). In computing the tunneling rate, we assume that the system is completely in state a at $t \to -\infty$; the tunneling rate is the probability of remaining in state a at $t \to \infty$, the end of dynamical evolution. At a fixed number of bosons, the dynamics of the quantized model (3.2) can be found by expanding a quantum state in terms of Fock states $|N_a, N_b\rangle$, where N_a and N_b are the number of particles in quantum states a and b, respectively.

Upon careful examination of the data in Fig. 3.1, one observes that $c = v$ is a critical value. When $c < v$, the tunneling rate goes to zero in the adiabatic limit $\alpha \to 0$ for both the mean-field model and the quantized model. However, when $c > v$, the tunneling rate for the mean-field model is always nonzero, while the tunneling rate can be zero for the quantized model. Since the mean-field model is the semiclassical limit of the quantized model, the mean-field result can be regarded as the result from the quantized model with the limit $N \to \infty$ having been taken. Therefore, the results in Fig. 3.1 show that the tunneling behavior in the quantized model (3.2) depends strongly on the order of the limits taken, while this dependence itself relies on the value of the mean-field interaction strength c [5].

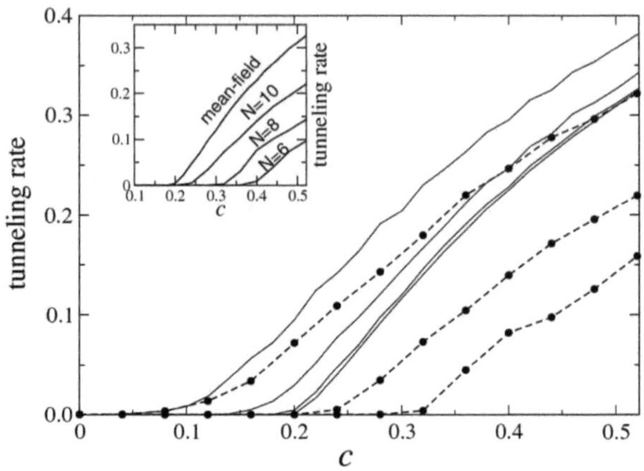

Fig. 3.1 Tunneling rate as a function of the mean-field interaction strength c for $v = 0.2$. The solid lines are obtained with the mean-field model (3.3) for $\alpha = 0.005, 0.001, 0.0001$, and 0.00001 (from top to bottom); the dot-dashed lines are obtained with the quantized model (3.2) for $\alpha = 0.001$, 0.0001, and 0.00001 (from top to bottom) with $N = 10$. The inset shows the tunneling rates for the mean-field model and the quantized model with different N at $\alpha = 0.0001$, demonstrating the convergence

3.1.4 Energy Band Structure

To understand the results shown in the section above, we first examine the energy levels of the second-quantized model (3.2) as functions of γ, the slowly changing system parameter. These energy levels, which can be found by directly diagonalizing the Hamiltonian $\hat{H}$, are plotted in Fig. 3.2. A sharp change appears in the structure of energy levels as the mean-field interaction c changes: a net of anticrossings appears in the lower part of the quantized energy levels when $c > v$. As described previously [8], when $c > v$, a loop structure emerges in the energy band of the mean-field model (3.3). When the mean-field energy levels (circles) are also plotted in Fig. 3.2, we find that the quantized energy levels are bounded by the mean-field energies. In particular, the mean-field energy levels envelop the net of anticrossings in the quantized energy levels. Such a correspondence was first reported in [4].

The structural change in the energy bands is associated with a change in the phase space of the mean-field model (3.3) as shown in Fig. 3.3. In plotting this figure, we note that the mean-field model, in fact, has only two independent variables and that its Hamiltonian can be reduced to [9]

$$H_{mf} = \gamma p - cp^2 + \frac{v}{2}\sqrt{1 - 4p^2}\cos q, \tag{3.4}$$

where $p = (|a|^2 - |b|^2)/2$ and $q = \theta_b - \theta_a$, with $\theta_{a,b}$ being the phases of a and b. It is clear from Fig. 3.3 that when $c < v$, there is one minimum and one maximum; when $c > v$, we observe two local minima, one maximum and one saddle point.

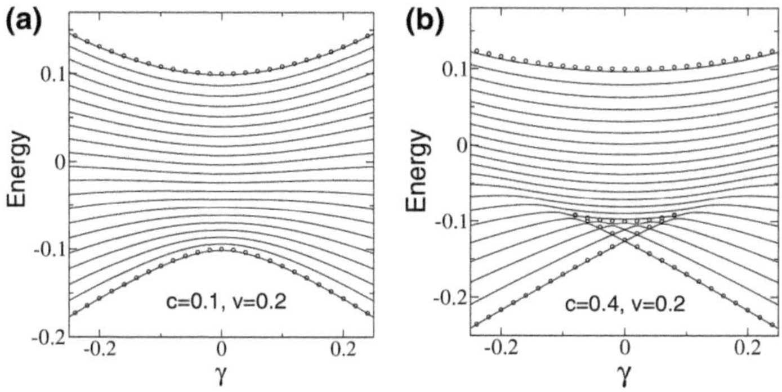

Fig. 3.2 Energy levels from the second-quantized model with $N = 20$ and the mean-field model. The solid lines are quantized energy levels; the open circles are mean-field energy levels. Note that for comparison with the mean-field theory, the quantized energy levels from $\hat{H}$ have been divided by N

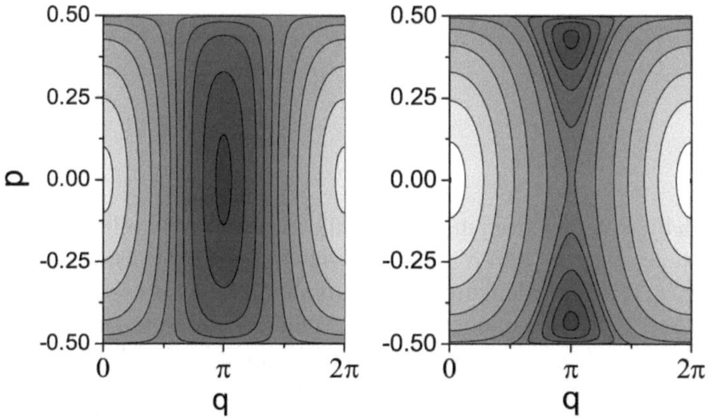

Fig. 3.3 Energy contours of the mean-field model H_{mf}. Left: $c = 0.1$, $v = 0.2$, and $\gamma = 0.0$; right: $c = 0.4$, $v = 0.2$, and $\gamma = 0.0$. The gray colors indicate the relative values of the mean-field energy, with darker gray for smaller energy values

Since these extremal points correspond to the eigenstates in the mean-field energy bands in Fig. 3.2 [9], we conclude that the structural change in the phase space is connected with the structural change in the energy levels.

This connection can be further explored by requantizing the mean-field model H_{mf} with the Sommerfeld theory, which states that the quantum motions are the periodic motions in the classical phase space that satisfy

$$\frac{1}{2\pi}\oint p dq = \frac{n\hbar}{N}, \quad n = 0, 1, 2, \ldots . \tag{3.5}$$

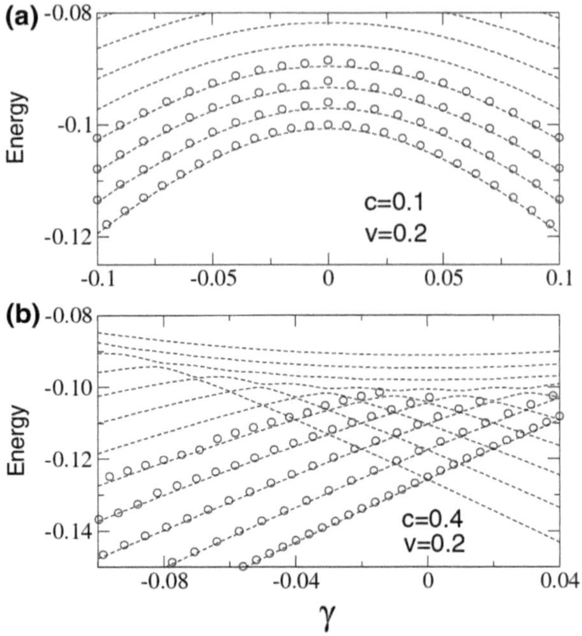

Fig. 3.4 Comparison between the energy levels of the second-quantized model (dashed lines) with $N = 40$ and the Sommerfeld energy levels (open circles). **a** $c = 0.1$, $v = 0.2$; **b** $c = 0.4$, $v = 0.2$. For clarity, we have plotted only a portion of the energy levels

The division by N comes from the fact that the mean-field Hamiltonian is an average for one particle, $H_{mf} = \langle \hat{H} \rangle / N$. One can view $\hbar_{eff} = \hbar/N$ as the effective Plank constant for H_{mf}. In our calculations, the natural unit $\hbar = 1$ is used. For convenience, we call the energy levels obtained from (3.5) the Sommerfeld energy levels. These levels are shown and compared with the quantized energy levels of $\hat{H}$ in Fig. 3.4.

When $c < v$, the mean-field Hamiltonian has exactly one maximum ($q = 0$) and one minimum ($q = \pi$). The Sommerfeld quantization around the maximum produces energy levels lower than the maximum energy, while the quantization around the minimum generates energy levels higher than the minimum. This condition explains why the mean-field energy levels bound the quantized energy levels in Fig. 3.2. We also observe that the energy gap arises from the different quantization number in (3.5), from which we estimate that the energy gap between the lowest two energy levels at $\gamma = 0$ is $\Delta \simeq v\sqrt{1 - c/v}$, independent of N, which is consistent with the numerical results shown in Fig. 3.5a.

When $c > v$, the phase space of H_{mf} becomes very different: two local minima exist with an additional saddle point. In this case, the Sommerfeld quantization around the two local minima gives rise to two sets of Sommerfeld energy levels. In the lower part of Fig. 3.4, for clarity, we have plotted only one set. If two sets were plotted, they would form a net of crossings, matching closely with the anticrossing net from $\hat{H}$. In implementing the Sommerfeld quantization, we have ignored the tunneling through the energy barrier between the two local minima. Once the tunneling is considered, degeneracies are lifted, and the crossings become anticrossings. This result shows that the energy gaps inside the triangular net have a different origin from the energy gaps

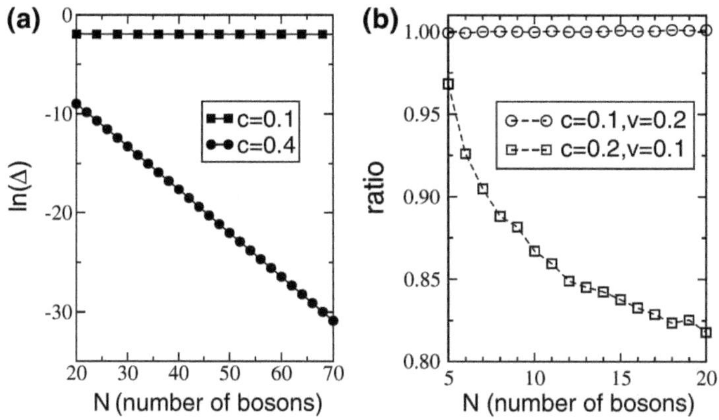

Fig. 3.5 **a** Energy gap between the lowest two eigenenergies in the second-quantized model at $\gamma = 0$. The squares represent $c = 0.1$, and the dots represent $c = 0.4$, with $v = 0.2$ for both. The solid line is the approximate result $\Delta = v\sqrt{1 - c^2/v^2}$ for $c < v$. **b** Ratio of the bosons in the right well at the end of the tunneling process. The computation is performed with the second-quantized model $\hat{H}$ with a sweeping rate $\alpha = 0.0001$

outside the net or in the case of $c < v$. The energy gaps produced at these crossings can be estimated with the WKB method. Since the effective Planck constant for H_{mf} is $\hbar/N$, we expect that the gaps decrease exponentially with N, which is exactly what the numerical results in Fig. 3.5a indicate [5].

3.1.5 *Commutability Between Two Limits*

It is now straightforward to understand the tunneling behavior shown in Fig. 3.1. Let us recall the Landau-Zener (LZ) tunneling in a two-level model [10]. As γ changes with time as $\gamma = \alpha t$, the LZ tunneling rate is $r_{LZ} = \mathrm{e}^{-\pi\Delta^2/2\alpha}$, where Δ is the energy gap between the two levels. For a multilevel system such as our second-quantized model $\hat{H}$, the above equation remains a satisfactory approximation for the tunneling rate between two consecutive energy levels. We use the tunneling between the two lowest energy levels as an example. As already analyzed, the energy gap changes with N as follows [5]:

$$\Delta = \begin{cases} \kappa_1 v, & c < v; \\ N\kappa_2 \mathrm{e}^{-\eta N}, & c > v. \end{cases} \tag{3.6}$$

The parameter $\kappa_1 \simeq \sqrt{1 - c/v}$. The other two parameters, κ_2 and η, can be computed with the WKB method as in [2] or with a more sophisticated method [11]. This approach leads to the following tunneling rate:

$$r \sim r_{LZ} = \begin{cases} \mathrm{e}^{-\frac{\pi\kappa_1^2 v^2}{2\alpha}}, & c < v; \\ \mathrm{e}^{-\frac{\pi N^2 \kappa_2^2}{2\alpha \mathrm{e}^{2\eta N}}}, & c > v. \end{cases} \tag{3.7}$$

For the case of $c < v$, it is clear that we have

$$\lim_{N\to\infty}\lim_{\alpha\to 0} r = \lim_{\alpha\to 0}\lim_{N\to\infty} r = 0, \tag{3.8}$$

which shows that the two limits $\alpha \to 0$ and $N \to \infty$ are commutable. This finding explains why when $c < v$, both sets of the tunneling rates in Fig. 3.1 become zero as $\alpha \to 0$.

For the other case, $c > v$, the tunneling rate takes different values at two different limits:

$$\begin{cases} \lim_{N\to\infty}\lim_{\alpha\to 0} r = 0; \\ \lim_{\alpha\to 0}\lim_{N\to\infty} r > 0. \end{cases} \tag{3.9}$$

This expression reveals that the two limits are no longer commutable. In the first limit, where the adiabatic limit $\alpha \to 0$ is taken at a fixed number of bosons, the energy gap is finite, and the system proceeds slowly enough not to cause tunneling. In the second limit, since the energy gap is already closed at $N \to \infty$, tunneling occurs regardless of how slowly γ changes, which explains why the tunneling rate from the mean-field model is always nonzero for $c > v$ (Fig. 3.5b). The incommutability of these two limits also implies that the mean-field theories, such as the GP equation for BECs, can be invalid for the adiabatic limit. One example is the Bloch states for a BEC in an optical lattice. In such a system, the Bloch wave number k can be regarded as an adiabatic parameter. If the GP equation was always valid in the adiabatic limit, it would mean that stable Bloch states should exist for all possible k. However, as shown in [12], a significant portion of Bloch states are unstable.

3.2 Quantum-Classical Correspondence of the Adiabatic Geometric Phase

3.2.1 *Interacting Bosonic Many-Body System*

We consider an interacting boson system described by the second-quantized Hamiltonian [13],

$$\hat{H} = \frac{Z}{2}(\hat{a}_1^\dagger \hat{a}_1 - \hat{a}_2^\dagger \hat{a}_2) + \frac{c}{4}(\hat{a}_1^\dagger \hat{a}_1 - \hat{a}_2^\dagger \hat{a}_2)^2 + \frac{\rho}{2}(\mathrm{e}^{i\phi}\hat{a}_1^\dagger \hat{a}_2 + \mathrm{e}^{-i\phi}\hat{a}_2^\dagger \hat{a}_1), \tag{3.10}$$

where $\hat{a}^\dagger$ ($\hat{a}$) and $\hat{b}^\dagger$ ($\hat{b}$) are the bosonic creation operators (annihilation operators) for two quantum states, labeled $|1\rangle$ and $|2\rangle$, respectively. Z is the energy offset between the two quantum states. The parameter $\rho e^{\pm i\phi}$ measures the coupling between the two atomic states, and $c = g/V$, with g being proportional to the atomic s-wave scattering length and V being the quantized volume of the atoms; $n = N/V$ is the mean density of the boson atoms, and the total number of atoms N is conserved. The above model can be derived from the bosonic-field Hamiltonian [14] and has received extraordinary attention in the literature on BECs. For $\phi = 0$, model (3.10) reduces to the well-known boson Josephson junction (BJJ) model [15, 16], which has been used to describe the tunneling dynamics of boson atoms between double-well potentials or two internal quantum states [3, 17].

The dynamics of system is governed by the Schrödinger equation, $i\hbar\partial|\Psi\rangle/\partial t = \hat{H}(\boldsymbol{R})|\Psi\rangle$, where $\boldsymbol{R} = (Z, \rho e^{\pm i\phi}, c)$ denotes the system parameters. The ith eigenstate $|\bar{\psi}_i(\boldsymbol{R})\rangle$ satisfies the eigenequation $\hat{H}(\boldsymbol{R})|\bar{\psi}_i(\boldsymbol{R})\rangle = E_i(\boldsymbol{R})|\bar{\psi}_i(\boldsymbol{R})\rangle$, with E_i being the eigenenergy. For convenience, we can expand $|\bar{\psi}_i(\boldsymbol{R})\rangle$ by Fock states; that is, $|\bar{\psi}_i(\boldsymbol{R})\rangle = \sum_{m=0}^{N} c_m^i(\boldsymbol{R})|m, N-m\rangle$, where c_m^i is the probability amplitude of the corresponding Fock state $|m, N-m\rangle = [(N-m)!m!]^{-1/2}(\hat{a}_1^\dagger)^m(\hat{a}_2^\dagger)^{N-m}|0\rangle$, with $m = 0, 1, \ldots, N$ and $N-m$ being the particle numbers in quantum states $|1\rangle$ and $|2\rangle$, respectively. The symbol $|0\rangle$ denotes the vacuum state, which is defined by $\hat{a}_1|0\rangle = \hat{a}_2|0\rangle = 0$. The Hamiltonian $\hat{H}(\boldsymbol{R})$ can thus be expressed as an $(N+1)\times(N+1)$ matrix, and the matrix element is given by $H_{m,n} = Z(2n-N)/2 \cdot \delta_{m,n} + c[n^2 + (N-n)^2 - 2n(N-n)]/4 \cdot \delta_{m,n} + \rho e^{i\phi}\sqrt{(N-n)(n+1)}/2 \cdot \delta_{m,n+1} + \rho e^{-i\phi}\sqrt{n(N-n+1)}/2 \cdot \delta_{m,n-1}$, with $\delta_{i,j}$ being the delta function. The coefficients $c_m^i(\boldsymbol{R})$ satisfy the normalization condition $\sum_m |c_m^i(\boldsymbol{R})|^2 = 1$. Even though these coefficients can be readily obtained by numerically diagonalizing the Hamiltonian matrix, they typically cannot be expressed in explicit forms analytically. However, with some approximations, we can derive their explicit forms [13].

3.2.2 *Mean-Field Hamiltonian*

For quantum many-body boson system (3.10), we define the trial state $|\Psi\rangle$ in terms of the SU(2) coherent states (CSs) [6], i.e., $|\Psi\rangle = \frac{1}{\sqrt{N!}}(\alpha\hat{a}_1^\dagger + \beta\hat{a}_2^\dagger)^N|0\rangle$, where $|0\rangle$ denotes the vacuum state as defined in the previous section. A variational scheme based on the above SU(2) CSs leads to the Hamiltonian [18]

$$\begin{aligned} H_{CS} = &\frac{NZ}{2}(|\alpha|^2 - |\beta|^2) + \frac{Nc_e}{4}(|\alpha|^2 - |\beta|^2)^2 \\ &+\frac{N\rho}{2}(e^{i\phi}\alpha^*\beta + e^{-i\phi}\alpha\beta^*), \end{aligned} \tag{3.11}$$

where α and β are complex probability amplitudes of states $|1\rangle$ and $|2\rangle$, respectively, that satisfy the normalization condition $|\alpha|^2 + |\beta|^2 = 1$. The parameter $c_e = cN$ denotes the effective atom-atom interactions.

According to [19], we define for convenience the total population probability $p_1 = |\alpha|^2 + |\beta|^2$ and the total phase $q_1 = \arg(\alpha)$; these two quantities form a pair of canonical variables. The other pair of canonical variables can be defined by the population probability $p_2 = |\beta|^2$ and the relative phase $q_2 = \arg(\alpha) - \arg(\beta)$. (Indeed, $p_2 = \alpha^2 - \beta^2$ and $q_2 = \arg(\beta) - \arg(\alpha)$ also form a pair of canonical variables, and this choice is equivalent to our choice; see [19] for details.) Using these canonical variables, a classical Hamiltonian can be obtained from (3.11):

$$\begin{aligned} H_C(p_1; p_2, q_2) = \frac{H_{CS}}{N} &= \frac{Z}{2}(p_1 - 2p_2) + \frac{c_e}{4}(p_1 - 2p_2)^2 \\ &+ \rho\sqrt{(p_1 - p_2)p_2}\cos(q_2 - \phi). \end{aligned} \tag{3.12}$$

The fixed points of the classical Hamiltonian are given by $p_1 = 1$ and $(\bar{p}_2, \bar{q}_2)$, which satisfy the following equations:

$$\left.\frac{\partial H_C(1; p_2, q_2)}{\partial p_2}\right|_{(\bar{p}_2, \bar{q}_2)} = \left.\frac{\partial H_C(1; p_2, q_2)}{\partial q_2}\right|_{(\bar{p}_2, \bar{q}_2)} = 0. \tag{3.13}$$

Following (3.13), we rewrite the fixed point equations as

$$c_e(1 - 2\bar{p}_2) + Z - \frac{(1 - 2\bar{p}_2)\rho\cos(\bar{q}_2 - \phi)}{\sqrt{1 - (1 - 2\bar{p}_2)^2}} = 0, \tag{3.14}$$

$$\frac{1}{2}\sqrt{1 - (1 - 2\bar{p}_2)^2}\rho\sin(\bar{q}_2 - \phi) = 0. \tag{3.15}$$

We treat Z as a perturbative parameter and expand the solution in the first order of Z around the fixed point for $Z = 0$, which takes the form

$$\bar{p}_2 = \bar{p}_2^{(0)} + \delta p \cdot Z, \tag{3.16}$$

where $\bar{p}_2^{(0)}$ is the solution of (3.14) for $Z = 0$. There are three solutions to $\bar{p}_2^{(0)}$, and we express them as follows:

$$\bar{p}_{2,0}^{(0)} = \frac{1}{2}, \quad \bar{p}_{2,\pm}^{(0)} = \frac{1}{2}\left(1 \pm \sqrt{1 - \frac{\rho^2}{c_e^2}}\right). \tag{3.17}$$

Substituting (3.16) back into (3.14) and neglecting the higher-order terms of Z, one can obtain the solutions of δp corresponding to the above three solutions of $\bar{p}_2^{(0)}$. Note that the change of the fixed points $(\bar{q}_2 = \phi) \to (\bar{q}_2 = \phi + \pi)$ is equivalent to the transformation $\rho \to -\rho$. We therefore only consider the situation $\rho > 0$ and $c_e < 0$. For $\bar{q}_2 = \phi + \pi$, the three solutions of δp are

$$\delta p_0 = \frac{1}{2(c_e + \rho)}, \quad \delta p_+ = \delta p_- = -\frac{\rho^2}{2c_e(c_e^2 - \rho^2)}. \tag{3.18}$$

Inserting these solutions into the classical Hamiltonian (3.12) and comparing the values of each $H_C(\bar{p}_2, \bar{q}_2)$, we obtain the fixed point that corresponds to the highest excited state (HES) [or the ground state (GS)] of the quantum system.

We now focus on the case of $c_e < 0$ and $\rho > 0$. Under this condition, when $|c_e| < \rho$, there are only two fixed points, and two more fixed points appear when $|c_e| > \rho$. The phase space of the above classical Hamiltonian for varied parameters is discussed in [9, 19]. When the parameter Z is small, we use the perturbation approach to calculate the fixed points analytically. By neglecting the higher-order terms of Z, for $\rho < |c_e|$, we have

$$(\bar{p}_2, \bar{q}_2) = \begin{cases} \left(\frac{1}{2}[1 \pm \sqrt{1 - \frac{\rho^2}{c_e^2}} - \frac{Z\rho^2}{c_e(c_e^2-\rho^2)}], \pi + \phi\right); \\ \left(\frac{1}{2}(1 + \frac{Z}{c_e - \rho}), \phi\right), \end{cases} \tag{3.19}$$

and for $\rho > |c_e|$,

$$(\bar{p}_2, \bar{q}_2) = \begin{cases} \left(\frac{1}{2}(1 + \frac{Z}{c_e + \rho}), \pi + \phi\right); \\ \left(\frac{1}{2}(1 + \frac{Z}{c_e - \rho}), \phi\right), \end{cases} \tag{3.20}$$

where the sign $\pm$ in the first line of (3.19) indicates the situations $Z > 0$ and $Z < 0$, respectively. From the perspective of classical-quantum correspondence, the two branches of fixed points (3.19) or (3.20) correspond to the GS and the HES of quantum model (3.10), respectively, because the fixed points correspond to the global minimum energy and the global maximum energy of the classical Hamiltonian [5].

We now give the explicit formula of the SU(2) CSs in terms of Fock states as follows [18]:

$$|\Psi\rangle = \frac{1}{\sqrt{N!}}(\alpha\hat{a}_1^\dagger + \beta\hat{a}_2^\dagger)^N|0\rangle = \sum_{m=0}^{N}\sqrt{\frac{N!}{(N-m)!m!}}\alpha^m\beta^{N-m}|m, N-m\rangle. \tag{3.21}$$

For the eigenstates $|\bar{\psi}_i(\boldsymbol{R})\rangle$, following (3.21) and using fixed points (3.19) and (3.20), the expressions of the coefficients $c_m^i(\boldsymbol{R})$ for the GS and the HES of quantum model (3.10) are given by

$$|c_m^i(\boldsymbol{R})|^2 = \frac{N!}{(N-m)!m!}(1-\bar{p}_2)^m\bar{p}_2^{N-m}. \tag{3.22}$$

We note that the above equation provides a convenient approximation of the quantum eigenstates when $|c_e| \leq 1$ and $|Z|\sqrt{N} \leq 1$. The analytical result (3.22) has been compared with the numerical results from direct diagonalization of the second-quantized Hamiltonian $\hat{H}(\boldsymbol{R})$. The agreement is satisfactory, as shown in Fig. 3.6 [13].

We now show the validity of the above CS description. When $\phi = 0$, (3.10) reduces to the model discussed by Leggett in a review article [3]. By defining the operators $\hat{J}_x \equiv (\hat{a}_1^\dagger\hat{a}_2 + \hat{a}_2^\dagger\hat{a}_1)/2$, $\hat{J}_y \equiv -i(\hat{a}_1^\dagger\hat{a}_2 - \hat{a}_2^\dagger\hat{a}_1)/2$, and $\hat{J}_z \equiv (\hat{a}_1^\dagger\hat{a}_1 - \hat{a}_2^\dagger\hat{a}_2)/2$,

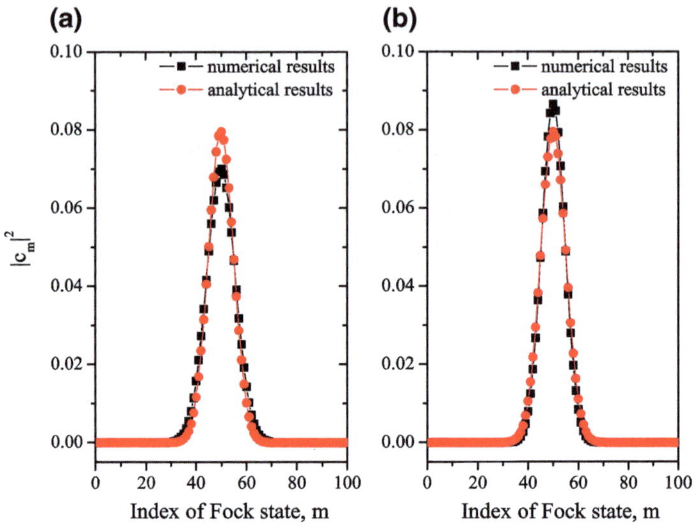

Fig. 3.6 Comparison between the numerical results from direct diagonalization of the second-quantized Hamiltonian $\hat{H}$ and the analytical solution given by (3.22) for the GS (**a**) and the HES (**b**). The parameters are $Z = 0.001$, $\rho = 0.5$, $N = 100$, and $c_e = -0.2$

the phase operator $\hat{\varphi}$ can be introduced as $\hat{E} \equiv e^{i\hat{\varphi}} = [(N/2 - \hat{J}_z)(N/2 + \hat{J}_z + 1)]^{-1/2}(\hat{J}_x + i\hat{J}_y)$. Then, the Hamiltonian can be rewritten in terms of these new operators as

$$\hat{H}(\hat{J}_z, \hat{\varphi}) = Z\hat{J}_z + c\hat{J}_z^2 + \rho\sqrt{\left(\frac{N}{2}\right)^2 - \hat{J}_z^2} \cos\hat{\varphi}. \tag{3.23}$$

The phase operators $\hat{\varphi}$ and $\hat{J}_z$ are canonical variables of the above system that satisfy

$$[\hat{J}_z, \hat{\varphi}] = -i. \tag{3.24}$$

It was proved in [3] that the state $|\Psi\rangle$ is a CS and has the following property:

$$\hat{E}|\Psi\rangle = e^{i\varphi}|\Psi\rangle + \delta \cdot |\Psi\rangle_\perp, \tag{3.25}$$

where φ is a real number, $|\Psi\rangle_\perp$ is a normalized state orthogonal to $|\Psi\rangle$, and δ is of order $N^{-1/2}$. Thus, one can prove that the uncertainty of the phase operator for state $|\Psi\rangle$ is of order $N^{-1/2}$, that is, $\Delta\hat{\varphi} \sim N^{-1/2}$.

Since state $|\Psi\rangle$ is a CS, the quantity $\Delta\hat{J}_z\Delta\hat{\varphi}$ for such a state has the following relation:

$$\Delta\hat{J}_z\Delta\hat{\varphi} \sim \mid \overline{[\hat{J}_z, \hat{\varphi}]} \mid . \tag{3.26}$$

It is straightforward to determine that $\Delta\hat{J}_z\Delta\hat{\varphi} \sim 1$ and therefore that the uncertainty of $\hat{J}_z$ for state $|\Psi\rangle$ is of order $N^{1/2}$, that is, $\Delta\hat{J}_z \sim N^{1/2}$. Then, we can estimate the uncertainty of $\hat{H}(\hat{J}_z, \hat{\varphi})$ for the CS $|\Psi\rangle$, which is

$$\Delta\hat{H} \sim Z\Delta\hat{J}_z + c(\Delta\hat{J}_z)^2 + \rho\Delta\hat{J}_z\Delta\hat{\varphi} \sim Z\sqrt{N} + cN + \rho. \tag{3.27}$$

For $|c|N = |c_e| \leq 1, |Z|\sqrt{N} \leq 1$, and $\rho \sim 1$, we observe that $\Delta\hat{H} \sim 1$. Considering that $\langle\Psi|\hat{H}|\Psi\rangle$ is of order N, one has $\Delta\hat{H}/\langle\Psi|\hat{H}|\Psi\rangle \sim 1/N$, which implies that if the total particle number $N \gg 1$, then (3.22) gives a satisfactory approximation of the quantum eigenstates of model (3.10).

3.2.3 *Quantum Berry Phase*

To calculate the Berry phase of the system, for simplicity, we fix the parameters Z, ρ, and c and change only the parameter ϕ adiabatically from 0 to 2π. We introduce the dimensionless adiabatic parameter $v \sim |d\phi/dt| \sim 1/T$ (T is the time duration) as the measure of how slow the parameter changes. The adiabatic parameter tends to zero, i.e., $v \to 0$, indicating the adiabatic limit. The quantum adiabatic theorem states that if the quantum system is initially prepared in an eigenstate $|\bar{\psi}_i(\boldsymbol{R}(0))\rangle$, then at time $t = T$, the system will be found in the eigenstate $|\bar{\psi}_i(\boldsymbol{R}(0))\rangle$, and a geometric phase factor will be acquired during the adiabatic process. This phase factor is independent of the time duration T and is related only to the geometric property of the closed path in parameter space. To calculate this geometric phase factor, we first make the variable transformation $\hat{a}_2' = \hat{a}_2 e^{i\phi}$ and substitute it back into (3.10); then, we obtain a new Hamiltonian $\hat{H}(\boldsymbol{R}')$ with a new parameter vector, $\boldsymbol{R}'(Z, \rho, c)$. $\hat{H}(\boldsymbol{R}')$ is not an explicit function of the time-dependent parameter $\phi(t)$. According to the Fock-state bases defined previously, we can rewrite the ith eigenstate of the system as $|\bar{\psi}_i(\boldsymbol{R})\rangle = \sum_{m=0}^{N} c_m^i(\boldsymbol{R}')e^{-i(N-m)\phi}|m, N-m\rangle$. Clearly, $c_m^i(\boldsymbol{R}) = c_m^i(\boldsymbol{R}')e^{-i(N-m)\phi}$, and $|c_m^i(\boldsymbol{R}')|^2 = |c_m^i(\boldsymbol{R})|^2$. Finally, for our interacting N-particle system, the Berry connection reads

$$\boldsymbol{A}_N(\boldsymbol{R}) = i\langle\bar{\psi}_i(\boldsymbol{R})|\nabla_{\boldsymbol{R}}|\bar{\psi}_i(\boldsymbol{R})\rangle = \sum_{m=0}^{N}(N-m)|c_m^i(\boldsymbol{R}')|^2. \tag{3.28}$$

The Berry phase for the ith eigenstate is the circuit integral of the Berry connection (3.28), that is,

$$\gamma_N = \oint \boldsymbol{A}_N(\boldsymbol{R}) \cdot d\boldsymbol{R} = 2\pi \sum_{m=0}^{N}(N-m)|c_m^i(\boldsymbol{R}')|^2. \tag{3.29}$$

It is clear that this quantum Berry phase is time independent and is explicitly N dependent. By applying the previous explicit expression of $|c_m^i(\boldsymbol{R})|^2$, for both the GS and the HES, we obtain the analytical results of γ_N for arbitrary particle number N.

3.2.4 Classical Hannay Angle

To find the classical analogy of the quantum Berry phase (3.29), we calculate the Hannay angle from the classical Hamiltonian H_C. We know that the classical Hannay angle determined by the evolution of angle variables can be obtained by making a canonical transformation to action-angle variables. This transformation can be achieved in terms of a generating function $S = \sum_i S_i(q_i, I_i; \phi(t))$ according to the scheme $\{p_i, q_i\} \rightarrow S_i(q_i, I_i; \phi(t)) \rightarrow \{\theta_i, I_i\}$ with $i = 1$ and 2. To find the canonical transformation between the old variables and the action-angle variables, we need to simplify the Hamiltonian H_C by expanding it around the fixed points.

We first consider the HES denoted by the fixed point $\bar{q}_2 = \phi$. Let us make a perturbation expansion around this fixed point by $p_2 = \bar{p}_2 + \tilde{p}$ and $q_2 = \bar{q}_2 + \tilde{q}$. Inserting these expansions into (3.12) and ignoring the terms of third order (i.e., $\tilde{p}^3$ and $\tilde{q}^3$) and higher, we obtain

$$H_D(p_1; p_2, q_2) = H_0(p_1) + T(p_1)\tilde{p} + G(p_1)\tilde{p}^2 + F(p_1)\tilde{q}^2, \tag{3.30}$$

where

$$H_0(p_1) = \frac{Z}{2}(p_1 - 2\bar{p}_2) + \frac{c_e}{4}(p_1 - 2\bar{p}_2)^2 + \rho\sqrt{(p_1 - \bar{p}_2)\bar{p}_2}, \tag{3.31}$$

$$T(p_1) = \frac{\rho}{2}\frac{(2\bar{p}_2 - 1)\sqrt{p_1 - \bar{p}_2} + (p_1 - 2\bar{p}_2)\sqrt{1 - \bar{p}_2}}{\sqrt{(p_1 - \bar{p}_2)(1 - \bar{p}_2)\bar{p}_2}} + c_e(1 - p_1), \tag{3.32}$$

$$G(p_1) = c_e - \frac{\rho p_1^2}{8[\bar{p}_2(p_1 - \bar{p}_2)]^{3/2}}, \tag{3.33}$$

$$F(p_1) = -\frac{\rho}{2}\sqrt{\bar{p}_2(p_1 - \bar{p}_2)}. \tag{3.34}$$

Considering $E = H_D(p_1; p_2, q_2)$ to be a constant, we have

$$p_1 = \Gamma(\tilde{p}, \tilde{q}; E) = l, \tag{3.35}$$

where $\Gamma(\tilde{p}, \tilde{q}; E)$ can be obtained by setting (3.30) as a constant E. Since p_1 is independent of $\tilde{p}$ and $\tilde{q}$, the above equation implies that l should be a constant, and then, we have

$$E - H_0(l) + \frac{T^2(l)}{4G(l)} = G(l)\left[\tilde{p} + \frac{T(l)}{2G(l)}\right]^2 + F(l)\tilde{q}^2. \tag{3.36}$$

For our system, one can assume that the generating function takes the form of $S(q_1, I_1, q_2, I_2; \phi(t)) = S_1(q_1, I_1; \phi(t)) + S_2(q_2, I_2; \phi(t))$, where S_1 and S_2 satisfy the equations $p_1 = \partial S_1(q_1, I_1; \phi(t))/\partial q_1$ and $p_2 = \partial S_2(q_2, I_2; \phi(t))/\partial q_2$, respectively. According to (3.35) and (3.36), we have

$$\frac{\partial S_1(q_1, I_1; \phi(t))}{\partial q_1} = l, \tag{3.37}$$

$$\frac{\partial S_2(q_2, I_2; \phi(t))}{\partial q_2} = \sqrt{\frac{E_h(l) - F(l)\tilde{q}^2}{G(l)}} + \bar{p}_2 - \frac{T(l)}{2G(l)}, \tag{3.38}$$

where $E_h(l) = E - H_0(l) + T^2(l)/(4G(l))$. By setting $p_h = \tilde{p} + T(l)/(2G(l))$, (3.36) becomes a harmonic-type Hamiltonian $E_h(l) = G(l)p_h^2 + F(l)\tilde{q}^2$, and we can define a new generating function by $p_h = \partial S_2'(\tilde{q}, I_2; \phi(t))/\partial \tilde{q} = \sqrt{(E_h(l) - F(l)\tilde{q}^2)/G(l)}$. With the help of the definition of actions, we have

$$I_1 = \frac{1}{2\pi} \oint \frac{\partial S_1(q_1, I_1; \phi(t))}{\partial q_1} dq_1 = l, \tag{3.39}$$

$$\begin{aligned} I_2 &= \frac{1}{\pi} \int_0^{\sqrt{E_h(l)/F(l)}} \frac{\partial S_2'(\tilde{q}, I_2; \phi(t))}{\partial \tilde{q}} d\tilde{q} \\ &= \frac{E_h(l)}{2\sqrt{G(l)F(l)}}. \end{aligned} \tag{3.40}$$

From (3.37) and (3.38), we can obtain the branches of the generating function by using the relation $p_2 = \partial S_2(q_2, I_2; \phi(t))/\partial q_2 = \partial S_2'(\tilde{q}, I_2; \phi(t))/\partial \tilde{q} + \bar{p}_2 - T(l)/(2G(l))$ and considering the fact that (p_2, q_2) are periodic evolutions around $(\bar{p}_2, \phi)$.

$$S_1(q_1, I_1; \phi(t)) = I_1 q_1, \tag{3.41}$$

$$S_2(q_2, I_2; \phi(t)) = S_g + \left[\bar{p}_2 - \frac{T}{2G}\right](q_2 - \phi), \tag{3.42}$$

where

$$S_g = \int_\phi^{q_2} \left[\frac{E_h(I_1) - F(I_1)(q_2 - \phi)^2}{G(I_1)}\right]^{1/2} dq_2 = 2I_2 \int_0^\Lambda \sqrt{1 - x^2} dx, \tag{3.43}$$

with $\Lambda = \sqrt{F(I_1)/E_h(I_1)}(q_2 - \phi)$. We can now express the generating function as

$$S(q_1, I_1, q_2, I_2; \phi(t)) = I_1 q_1 + 2I_2 \int_0^\Lambda \sqrt{1 - x^2} dx + \left[\bar{p}_2 - \frac{T}{2G}\right](q_2 - \phi). \tag{3.44}$$

Finally, through a canonical transformation, we obtain a new Hamiltonian, $\bar{H}(I_1, \theta_1, I_2, \theta_2; t)$, that differs from the old Hamiltonian $H_D(p_1; p_2, q_2)$ both in value and in functional form since the canonical transformation is time dependent through the slowly changing parameter $\phi(t)$. In fact,

$$\bar{H}(I_1, \theta_1, I_2, \theta_2; t) = H_I(I_1, I_2; \phi) + \frac{d\phi}{dt}\frac{\partial S(q_1, I_1, q_2, I_2; \phi)}{\partial \phi}, \tag{3.45}$$

where $H_I(I_1, I_2; \phi) \equiv H_D(p_1; p_2, q_2)$, which is the angle-independent Hamiltonian related to the parameter ϕ, takes the form

$$H_I(I_1, I_2; \phi(t)) = 2I_2\sqrt{G(I_1)F(I_1)} + H_0(I_1) - \frac{T^2(I_1)}{4G(I_1)}. \tag{3.46}$$

The Hamilton's equation for the angle variables is $d\theta_i/dt = \partial \bar{H}/\partial I_i$, with $i = 1$ and 2. We apply this equation to (3.45); the first term gives the part of the evolution that would occur even if the parameters remained constant, arising from the frequencies $\omega_i(I_i; \phi) = \partial H_I(I_1, I_2; \phi)/\partial I_i$, with $i = 1$ and 2. Furthermore, from the second term in (3.45), we can obtain the Hannay angle by

$$\theta_H^i = \int_0^T dt \frac{d\phi}{dt}\frac{\partial}{\partial I_i}\left(\frac{\partial S}{\partial \phi}\right). \tag{3.47}$$

Note that the Berry phase is the shift of the total phase; thus, this phase corresponds to the Hannay angle for the action I_1. From (3.44), we have $\partial(\partial S/\partial\phi)/\partial I_1 = \partial(\partial S_g/\partial\phi)/\partial I_1 + \partial[T(I_1)/(2G(I_1))]/\partial I_1$. Note that the term of $\partial(\partial S_g/\partial\phi)/\partial I_1$ in (3.47) is proportional to the action I_2; therefore, this term makes no contribution to the Hannay angle for I_1. By using the explicit expressions of (3.32) and (3.33) and then considering $I_1 = 1$ and $I_2 = 0$ for the eigenstates, a straightforward but tedious calculation gives the Hannay angle for the HES as follows:

$$\theta_H^{HES} = \oint \frac{\rho\bar{p}_2 - 4c_e[(1-\bar{p}_2)\bar{p}_2]^{3/2}}{-\rho + 8c_e[(1-\bar{p}_2)\bar{p}_2]^{3/2}} d\phi. \tag{3.48}$$

Similarly, we can obtain the Hannay angle for the GS (i.e., $\bar{q}_2 = \pi + \phi$) as well; this expression takes the form

$$\theta_H^{GS} = \oint \frac{-\rho\bar{p}_2 - 4c_e[(1-\bar{p}_2)\bar{p}_2]^{3/2}}{\rho + 8c_e[(1-\bar{p}_2)\bar{p}_2]^{3/2}} d\phi. \tag{3.49}$$

To express the above integrals in explicit form, we substitute (3.19) into (3.48) and (3.49); keeping the terms of first order in parameter Z, we have

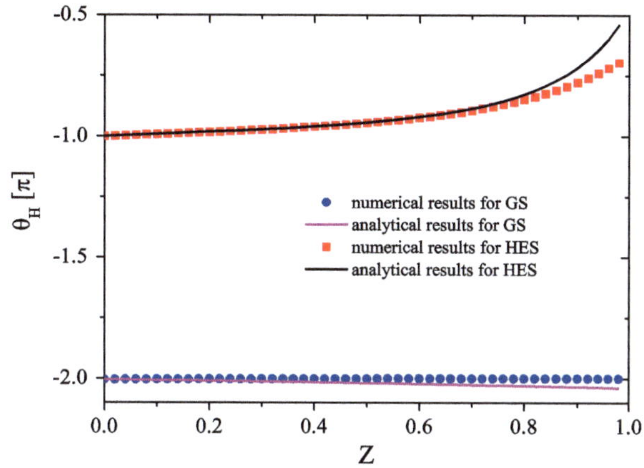

Fig. 3.7 Comparison between the approximate analytical solutions and the exact numerical solutions of the Hannay angle for both the GS and the HES.The parameters are $\rho = 0.1$ and $c_e = -1$

$$\theta_H^{HES} = -\pi \left[1 - \frac{Z\rho}{(c_e - \rho)^2} \right], \tag{3.50}$$

$$\theta_H^{GS} = -\pi \left[1 - \frac{2c_e Z\rho^2}{(c_e^2 - \rho^2)^2} \mp \frac{c_e\sqrt{c_e^2 - \rho^2}}{c_e^2 - \rho^2} \right]. \tag{3.51}$$

To check the above two approximate analytical results, we solve (3.13) numerically; the results are shown in Fig. 3.7. The analytical results are suitably accurate in the small-Z region [13].

3.2.5 Connection Between the Berry Phase and the Hannay Angle

With the explicit expressions of the Berry phase and Hannay angle at hand, we can determine the relation between them. We first calculate the number derivative of the quantum Berry phase. Substituting our previously obtained analytical solutions (3.22) into (3.29), we can obtain the particle number derivative of the quantum Berry phase analytically for the GS,

$$\frac{\partial \gamma_N^{GS}}{\partial N} = \pi \left[1 - \frac{2c_e Z\rho^2}{(c_e^2 - \rho^2)^2} \mp \frac{c_e\sqrt{c_e^2 - \rho^2}}{c_e^2 - \rho^2} \right], \tag{3.52}$$

and for the HES,

$$\frac{\partial \gamma_N^{HES}}{\partial N} = \pi \left[1 - \frac{Z\rho}{(c_e - \rho)^2} \right]. \tag{3.53}$$

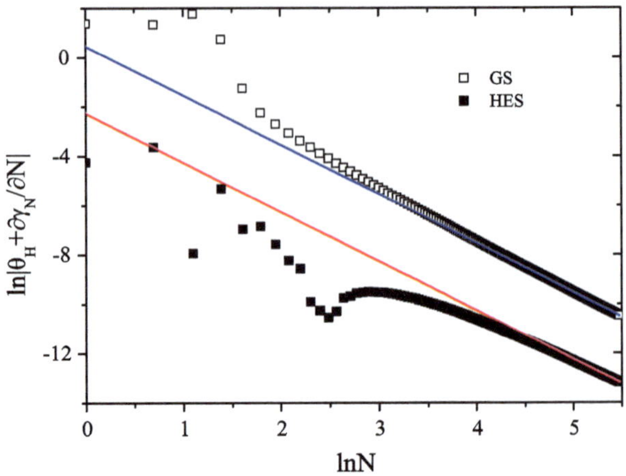

Fig. 3.8 Difference between the number derivative of the Berry phase and the minus Hannay angle versus the particle number for both the GS and the HES. All results shown here use the exact numerical solutions. Solid lines are plotted to guide the eye, with the slope being −2. The parameters are $Z = 0.01$, $c = -0.1$, and $\rho = 0.1$

Comparing these results with the Hannay angles given by (3.31) and (3.32), we notably find that they are identical except for a sign difference. Even though the analytical relation shows that $\partial\gamma_N/\partial N$ exactly equals $-\theta_H$, there indeed exists a small difference proportional to the quantum fluctuation of order $(1/N)$ between the two quantities because of the approximation that was used in deriving the analytical expressions of the quantum eigenstates.

We have numerically calculated the quantum eigenstates (i.e., c_m^i) and the classical fixed points [i.e., $(\bar{p}_2, \bar{q}_2)$] and obtained the exact numerical values of $\partial\gamma_N/\partial N$ and $-\theta_H$ for both the GS and the HES. The results show that the difference between the minus classical Hannay angle and the particle number derivative of the quantum Berry phase tends to 0 with increasing particle number N. Moreover, their differences converge to 0 in the power law of $1/N^2$ (see Fig. 3.8). Therefore, we present the relation between the classical Hannay angle and the quantum Berry phase as follows [13]:

$$\theta_H = -\frac{\partial\gamma_N}{\partial N} + O(N^{-2}). \tag{3.54}$$

This formula is the key result of the present section; we now provide some explanation. According to adiabatic theory, we know that an N-body wave function acquires a total phase $\lambda_N = \int E_N dt + \gamma_N$ during a cyclically adiabatic evolution. In the SU(2) coherent-state description, the corresponding single-particle wave function has a phase shift $\lambda = \partial\lambda_N/\partial N = \int \mu dt + \partial\gamma_N/\partial N$, where $\mu = \partial E_N/\partial N$ is the chemical potential; its time integral gives the dynamic phase. The second term, $\partial\gamma_N/\partial N$, characterizes the geometric change and corresponds to the classical Hannay angle θ_H.

Our relation $\theta_H \sim -\partial\gamma_N/\partial N$ is similar to the well-known relation established by Berry at the semiclassical level [20], where the classical action has been replaced by the total particle number in our many-body case. For our many-body system, the total

Table 3.1 Comparison of the adiabatic geometric properties with different descriptions for the interacting many-body bosonic system

System	Interacting many-body bosonic system		
Description	Second-quantized model: $\hat{H} = \hat{H}(\hat{a}_j, \hat{a}_j^\dagger; \boldsymbol{R})$, $\hat{a}_j^\dagger$ ($\hat{a}_j$) is the bosonic creation (annihilation) operator, $\boldsymbol{R}$ is the parameter vector	Mean-field model: $H_{MF}(\boldsymbol{R}) = \lim_{N\to\infty} \frac{\langle\Psi\|\hat{H}\|\Psi\rangle}{N}$, with $n = \frac{N}{V} =$ constant, $\|\Psi\rangle = \frac{1}{\sqrt{N!}}(\sum_j \alpha_j \hat{a}_j^\dagger)^N \|0\rangle$ is the SU(2) coherent state, V is the quantized volume	Classical Hamiltonian: $H_C(N; \boldsymbol{R}) = \frac{\langle\Psi\|\hat{H}\|\Psi\rangle}{N}$, N is the total number of particles, $\|\Psi\rangle$ is the SU(2) coherent state
Adiabatic geometric property	Quantum Berry phase: $\gamma_N = \oint \boldsymbol{A}_N(\boldsymbol{R}) \cdot d\boldsymbol{R}$, $\boldsymbol{A}_N(\boldsymbol{R}) = i\langle\bar{\psi}\|\nabla_{\boldsymbol{R}}\|\bar{\psi}\rangle$ is the Berry connection, $\|\bar{\psi}(\boldsymbol{R})\rangle$ is the eigenstate	Nonlinear geometric phase: $\gamma_G = \gamma - \gamma_D = \gamma_B + \gamma_{NL}$, γ is the total phase and $\gamma_D = \int \mu(\boldsymbol{R})dt$ is the dynamical phase with μ being the chemical potential. γ_{NL} is the nonlinear correction, Berry phase is $\gamma_B = i\oint \bar{\psi}^* \nabla_{\boldsymbol{R}} \bar{\psi} \cdot d\boldsymbol{R}$ with $\bar{\psi}$ being the eigenstate in projective Hilbert space	Classical Hannay angle: $\theta_H = \int_0^T dt \frac{d\boldsymbol{R}}{dt} \frac{\partial}{\partial I} \left(\frac{\partial S}{\partial \boldsymbol{R}}\right)$, I is the action variable, S is the generating function, T is the period of the cycle evolution
Connection	For $N \gg 1$, $\theta_H = -\frac{\partial \gamma_N}{\partial N} + O(N^{-2})$; for $N \to \infty$ and $n = \frac{N}{V} =$ const., $\theta_H = -\gamma_G$		

particle number N together with the total phase exactly forms a pair of canonical conjugate variables, and the number of particles is naturally quantized as an integer.

We add two remarks. First, we note that the classical Hamiltonian H_C reduces to the mean-field Hamiltonian in the thermodynamic limit $N \to \infty$, while the particle density $n = N/V$ is kept constant [5]. In this limit, we compare our classical Hannay angle with the mean-field Berry phase obtained in [21] and find that they are identical except for a sign difference (see Table 3.1). Second, in practical experiments, the single-particle phase is expected to affect the interference of the condensates and can be measured in BEC systems [22]. From this perspective, we anticipate that our classical Hannay angle will be detected in experiments in many-body systems because this angle is just the single-particle phase change during the adiabatic process in the thermodynamic limit.

References

1. Hwang, J.T., Pechukas, P.: J. Chem. Phys. **67**, 4640 (1977)
2. Berry, M.V.: J. Phys. A **17**, 1225 (1984)
3. Leggett, A.J.: Rev. Mod. Phys. **73**, 307 (2001)
4. Karkuszewski, Z.P., Sacha, K., Smerzi, A.: Eur. Phys. J. D **21**, 251 (2002)
5. Wu, B., Liu, J.: Phys. Rev. Lett. **96**, 020405 (2006)
6. Yaffe, L.G.: Rev. Mod. Phys. 54, 407 (1982); Zhang, W.M., Feng, D.H., Gilmore, R. Rev. Mod. Phys. **62**, 867 (1990)
7. Inouye, S.: Nature (London) **392**, 151 (1998); Courteille, P., et al.: Phys. Rev. Lett. **81**, 69 (1998)
8. Wu, B., Niu, Q.: Phys. Rev. A **61**, 023402 (2000)
9. Liu, J., Fu, L.B., Ou, B.Y., Chen, S.G., Choi, D.I., Wu, B., Niu, Q.: Phys. Rev. A **66**, 023404 (2002)
10. Zener, C.: Proc. R. Soc. A **137**, 696 (1932)
11. Garg, A.: Phys. Rev. B **64**, 094413 (2001). Phys. Rev. B **64**, 094414 (2001)
12. Wu, B., Niu, Q.: Phys. Rev. A **64**, 061603 (2001)
13. Li, S.C., Liu, J., Fu, L.B.: Phys. Rev. A **83**, 042107 (2011)
14. Cirac, J.I., Lewenstein, M., Mølmer, K., Zoller, P.: Phys. Rev. A **57**, 1208 (1998)
15. Gati, R., Oberthaler, M.K.: J. Phys. B **40**, R61 (2007)
16. Chuchem, M., Smith-Mannschott, K., Hiller. M., Kottos, T., Vardi, A., Cohen, D.: Phys. Rev. A **82**, 053617 (2010)
17. Karkuszewski, Z.P., Sacha, K., Smerzi, A.: Eur. Phys. J. D **21**, 251 (2002)
18. Buonsante, P., Penna, V.: J. Phys. A **41**, 175301 (2008)
19. Liu, J., Wu, B., Niu, Q.: Phys. Rev. Lett. **90**, 170404 (2003)
20. Berry, M.V.: J. Phys. A **18**, 15 (1985)
21. Liu, J., Fu, L.B.: Phys. Rev. A **81**, 052112 (2010)
22. Andrews, M.R., Townsend, C.G., Miesner, H.-J., Durfee, D.S., Kurn, D.M., Ketterle, W.: Science **273**, 637 (1997)

Chapter 4
Exotic Virtual Magnetic Monopoles and Fields

Abstract In this chapter, we introduce exotic virtual magnetic monopoles and fields; these phenomena are formed in two typical systems. The disk-shaped virtual magnetic field is found in a long-range interacting spin-half system with a mean-field description. The fractional virtual magnetic monopole occurs in a mean-field ultracold atom-molecule conversion system. The virtual magnetic monopole chain is demonstrated in a quantum many-body ultracold atom-molecule conversion system.

4.1 Disk-Shaped Virtual Magnetic Field

In this section, we analytically calculate the Berry phase of a long-range interacting spin-$1/2$ system in the mean-field perspective. This mean-field Berry phase reduces to the Berry phase of the single spin-$1/2$ system when the interaction vanishes. The magnetic-like flux interpretation of the Berry phase shows that the source and sink of the relevant magnetic-like field are located, respectively, at the disk-shaped level-crossing region, where the first-order quantum phase transition occurs, and its boundary, where the continuous quantum phase transition occurs. Specifically, one part of the field originates from the level-crossing region and ends at its boundary, while the other part comes from infinity and also ends at its boundary. The shape of the interface between these two parts reflects the critical property of the system. From the asymptotic distribution of the field at infinity, we find that the source and sink as a whole can be interpreted as a disk-shaped monopole with a negative magnetic charge. We emphasize that all these results generalize the abovementioned classic paradigm to the interacting spin-$1/2$ system and that the analogues of the disk-shaped monopole and its exotic magnetic-like field can generally exist in other interacting spin systems [1].

We consider a long-range interacting spin-$1/2$ system in an adjustable external magnetic field $\boldsymbol{R} = (x, y, z) = (\rho \cos\phi, \rho \sin\phi, z)$. In appropriate units, the corresponding Hamiltonian takes the form

J. Liu et al., *Nonlinear Adiabatic Evolution of Quantum Systems*,
https://doi.org/10.1007/978-981-13-2643-1_4

$$H = \sum_{i=1}^{N} \boldsymbol{R} \cdot \boldsymbol{\sigma}_i + \frac{K}{N} \sum_{i \neq j=1}^{N} \sigma_i^z \sigma_j^z, \tag{4.1}$$

where $\boldsymbol{\sigma}_i = (\sigma_i^x, \sigma_i^y, \sigma_i^z)$ is the Pauli matrix vector of the ith spin, K is the reduced interaction constant, and N ($\gg 1$) is the spin number. Expressing the state of the system as $|\Psi\rangle = |\psi\rangle^N$ and applying the mean-field approximation to the Hamiltonian (4.1), we obtain the mean-field Hamiltonian $H_m = \boldsymbol{G} \cdot \boldsymbol{\sigma}$, where $\boldsymbol{G} = \boldsymbol{R} + (0, 0, K\langle\psi|\sigma^z|\psi\rangle)$. Then, the Schrödinger equation $id|\Psi\rangle/dt = H|\Psi\rangle$ (with $\hbar = 1$) reduces to the mean-field Schrödinger equation $id|\psi\rangle/dt = H_m|\psi\rangle$. Inserting $|\psi\rangle = \mathrm{e}^{-i\lambda}(\cos\frac{\alpha}{2}, \sin\frac{\alpha}{2}\mathrm{e}^{i\beta})^T$ into this equation yields

$$\frac{d\alpha}{dt} = -2\rho\sin(\beta - \phi), \tag{4.2}$$

$$\frac{d\beta}{dt} = 2z + 2K\cos\alpha - 2\rho\cot\alpha\cos(\beta - \phi), \tag{4.3}$$

$$\frac{d\lambda}{dt} = z + K\cos\alpha + \rho\tan\frac{\alpha}{2}\cos(\beta - \phi). \tag{4.4}$$

For a fixed K, (4.2) and (4.3) connect the projective Hilbert space spanned by (α, β) with the parameter space spanned by (ρ, ϕ, z). If we introduce the mean spin vector $\boldsymbol{s} = (s_x, s_y, s_z) = \langle\psi|\boldsymbol{\sigma}|\psi\rangle = (\sin\alpha\cos\beta, \sin\alpha\sin\beta, \cos\alpha)$, (4.2) and (4.3) can be expressed in the following compact form:

$$\frac{d\boldsymbol{s}}{dt} = 2\boldsymbol{G} \times \boldsymbol{s}. \tag{4.5}$$

We now assume that the point $(\bar{\alpha}, \bar{\beta})$ is a fixed point of the projective Hilbert space, i.e., $d\bar{\alpha}/dt = 0$, and $d\bar{\beta}/dt = 0$. At this point, (4.2) and (4.3), respectively, become

$$\bar{\beta} = \phi \ \text{ or } \ \bar{\beta} = \phi + \pi, \tag{4.6}$$

$$z + K\cos\bar{\alpha} = \pm\rho\cot\bar{\alpha}, \tag{4.7}$$

where the upper sign corresponds to $\bar{\beta} = \phi$ and the lower sign corresponds to $\bar{\beta} = \phi + \pi$. Similarly, (4.5) becomes $\bar{\boldsymbol{G}} \times \bar{\boldsymbol{s}} = 0$, where $\bar{\boldsymbol{s}} = (\bar{s}_x, \bar{s}_y, \bar{s}_z) = (\sin\bar{\alpha}\cos\bar{\beta}, \sin\bar{\alpha}\sin\bar{\beta}, \cos\bar{\alpha})$ and $\bar{\boldsymbol{G}} = \boldsymbol{R} + (0, 0, K\cos\bar{\alpha})$. Because the fixed point of the projective Hilbert space corresponds to the eigenstate of H_m, (4.6) and (4.7) actually determine all eigenstates of H_m. Furthermore, from (4.6) and (4.7), we find that the eigenvalue corresponding to the eigenstate $|\bar{\psi}\rangle = (\cos\frac{\bar{\alpha}}{2}, \sin\frac{\bar{\alpha}}{2}\mathrm{e}^{i\bar{\beta}})^T$ can be expressed as

$$\mu = \langle\bar{\psi}|H_m|\bar{\psi}\rangle = \pm\frac{\rho}{\sin\bar{\alpha}}, \tag{4.8}$$

where the sign convention is the same as in (4.7).

Table 4.1 Eigenvalues and eigenstates of H_m when $z = 0$

Eigenstate	$\bar{s}_z = \cos\bar{\alpha}$	$\bar{\beta}(K > 0)$	$\bar{\beta}(K < 0)$	μ
$\|1\rangle$	0	ϕ	ϕ	ρ
$\|2\rangle$	0	$\phi + \pi$	$\phi + \pi$	$-\rho$
$\|3\rangle$	$-\sqrt{1-\rho^2/K^2}$	ϕ	$\phi + \pi$	K
$\|4\rangle$	$\sqrt{1-\rho^2/K^2}$	ϕ	$\phi + \pi$	K

When $z = 0$, the eigenstates and eigenvalues of H_m are shown in Table 4.1. Here, we stress that the eigenstates $|3\rangle$ and $|4\rangle$ exist only when $\rho < |K|$. From Table 4.1, we can find the ground state, i.e., the eigenstate with the minimum μ. For the ferromagnetic interaction case where $K < 0$, the disk-shaped region determined by $z = 0$ and $\rho < |K|$ is the level-crossing region where the eigenstates $|3\rangle$ and $|4\rangle$ serve as the degenerate ground states and the first-order quantum phase transition occurs [2, 3]. Outside this region, the eigenstate $|2\rangle$ serves as the ground state instead of eigenstates $|3\rangle$ and $|4\rangle$. When ρ crosses the boundary of this region determined by $z = 0$ and $\rho = |K|$ from the outside to the inside, $\bar{s}_z$ of the ground state changes continuously from zero to a nonzero value and can thus be used as the order parameter to indicate that a continuous quantum phase transition occurs at the boundary [1]. Actually, similar quantum phase transitions occur widely in interacting spin systems [2–4]. When $z \neq 0$, the eigenstates and eigenvalues of H_m can be numerically obtained from (4.6), (4.7), and (4.8), and no other quantum phase transition is found. In general, we can analytically prove that H_m has two eigenstates when $K^{2/3} < \rho^{2/3} + z^{2/3}$ and has four eigenstates when $K^{2/3} > \rho^{2/3} + z^{2/3}$.

We now consider that the external field $\boldsymbol{R}$ changes with time, and we use the dimensionless adiabatic parameter $\varepsilon \sim |d\boldsymbol{R}/dt|$ to measure its rate of change. Furthermore, we assume that ε is small enough that—according to the adiabatic evolution condition in mean-field models [5, 6]—the system, which is initially in an eigenstate of H_m, can remain in this eigenstate and can thus evolve simultaneously with $\boldsymbol{R}$. When $\boldsymbol{R}$ returns to its initial value, the system acquires a mean-field Berry phase γ. Here, we note that because H_m includes α and is thus state dependent, γ cannot be expressed in the conventional form, i.e., $\gamma \neq -i\oint_L \langle\bar{\psi}|\nabla|\bar{\psi}\rangle \cdot d\boldsymbol{R} = \frac{1}{2}\oint_L (1-\cos\bar{\alpha})d\phi$, where L is the evolution loop of the system in the parameter space. To obtain the expression for γ, we use the method introduced in [7] to separate the γ-related term from the expression for $d\lambda/dt$. To proceed, we first note that because the adiabatic parameter ε is small but finite, the system fluctuates around the eigenstate during the evolution, which implies that $\alpha = \bar{\alpha} + \delta\alpha$ and $\beta = \bar{\beta} + \delta\beta$, where $\delta\alpha \sim \delta\beta \sim O(\varepsilon)$. Then, from (4.4), (4.7), and (4.8), we obtain

$$\frac{d\lambda}{dt} = \mu + \left[\frac{\mu}{2\cos^2(\bar{\alpha}/2)} - K\right]\sin\bar{\alpha}\delta\alpha + O(\varepsilon^2), \tag{4.9}$$

where the zero-order term μ has been completely decoupled. Integrating this term over the evolution time, we obtain the corresponding dynamical phase. Furthermore, from (4.3), (4.7), and (4.8), we obtain

$$\frac{d\bar{\beta}}{dt} = \frac{2\mu - 2K\sin^2\bar{\alpha}}{\sin\bar{\alpha}}\delta\alpha + O(\varepsilon^2). \tag{4.10}$$

Here, we note that $d\delta\beta/dt \sim O(\varepsilon^2)$. Combining (4.9) and (4.10) yields

$$\frac{d\lambda}{dt} = \mu + \frac{1}{2}\left[1 - \frac{\cos\bar{\alpha}}{1-(K/\mu)\sin^2\bar{\alpha}}\right]\frac{d\bar{\beta}}{dt} + O(\varepsilon^2), \tag{4.11}$$

where the first-order term, i.e., the γ-related term, has been completely decoupled. Integrating this term over the evolution period and using (4.6), we find that the mean-field Berry phase is

$$\gamma = \frac{1}{2}\oint_L\left[1 - \frac{\cos\bar{\alpha}}{1-(K/\mu)\sin^2\bar{\alpha}}\right]d\phi. \tag{4.12}$$

Since the above derivation does not involve any restrictions on the eigenstate or the evolution loop L, (4.12) is a general analytical expression for the mean-field Berry phase of the system. In contrast to the previous result [8], (4.12) includes the accumulative effect of the fluctuation during the evolution.

When $K = 0$, i.e., the interaction vanishes, (4.12) clearly reduces to the expression for the Berry phase of the single spin-1/2 system. When $K \neq 0$, the accuracy of (4.12) can be confirmed by a numerical calculation if we consider γ as the difference between the total phase and the dynamical phase in the adiabatic limit. For simplicity, we assume that $\phi = 2\pi t/T$ with both ρ and z fixed. Under this assumption, the numerical calculation consists of the following steps: (i) solve $\bar{\beta}$ and $\bar{\alpha}$ from (4.6) and (4.7), and substitute $\bar{\alpha}$ into (4.8) to obtain μ; (ii) integrate (4.2), (4.3), and (4.4) from 0 to T with the initial values $\alpha_0 = \bar{\alpha}$, $\beta_0 = \bar{\beta}$ and $\lambda_0 = 0$ to obtain λ; and (iii) compare the result of (4.12) with the value that $(\lambda - \mu T)$ approaches at large T.

The integrand in (4.12) diverges when

$$\mu = K\sin^2\bar{\alpha}. \tag{4.13}$$

From (4.7), (4.8), and (4.13), we obtain that

$$K^{2/3} = \rho^{2/3} + z^{2/3}, \tag{4.14}$$

$$\mu = K^{1/3}\rho^{2/3}, \tag{4.15}$$

$$\cos\bar{\alpha} = -z^{1/3}/K^{1/3}, \tag{4.16}$$

where (4.14) determines the divergence-related region in the parameter space and (4.15) and (4.16) determine the divergence-related μ and eigenstate as functions of the parameters. Here, we emphasize that (4.14), (4.15), and (4.16) apply to all eigenstates of H_m. From the previous description about eigenstates, we know that the divergence-related region determined by (4.14) is exactly the region where the number of the eigenstates changes, which indicates that the behavior of γ accurately reflects the properties of the eigenstate.

From the derivation of (4.12), we conclude that the Berry phase γ is closely related to the fluctuation $\delta\alpha$. Substituting (4.6) and (4.13) into (4.10), we find that at the divergence-related region, a slow change of ϕ leads to an infinite fluctuation $\delta\alpha$. For the present mean-field model, an infinite fluctuation around the ground state arises only when the continuous quantum phase transition occurs. Therefore, the divergence of the ground-state Berry phase γ_g can be interpreted as the consequence of the quantum criticality. Here and below, the subscript g refers to the ground state.

From the form of (4.12), we find that the mean-field Berry phase γ is not proportional to any solid angle and thus has no solid angle interpretation. Nevertheless, we can always interpret γ as the flux of a magnetic-like field in the parameter space as long as the field takes the appropriate form. This magnetic-like flux interpretation actually serves as the differential formulation for γ. In the following, we perform the interpretation of the ground-state Berry phase γ_g and reveal the relation between this interpretation and the quantum phase transitions of the system.

We first define the vector potential $\boldsymbol{A}_g$ satisfying $\oint_L \boldsymbol{A}_g \cdot d\boldsymbol{R} = \gamma_g$. Because $\bar{\beta}_g = \phi + \pi$, we have

$$\boldsymbol{A}_g = \left(\frac{1}{2\rho} - \frac{\cos\bar{\alpha}_g}{2\rho + 2K\sin^3\bar{\alpha}_g}\right)\hat{\boldsymbol{e}}_\phi, \tag{4.17}$$

where $\hat{\boldsymbol{e}}_j$ denotes the unit vector in direction j, with $j = \rho$, z, and ϕ. We emphasize that $\boldsymbol{A}_g$ is independent of the evolution loop L. We then define the magnetic-like field $\boldsymbol{B}_g$ satisfying $\boldsymbol{B}_g = \nabla \times \boldsymbol{A}_g - \boldsymbol{\delta}_g$, where $\boldsymbol{\delta}_g$ denotes the contribution from the Dirac string. A direct calculation gives

$$\boldsymbol{B}_g = \left(U\frac{\partial\bar{\alpha}_g}{\partial\rho} + V\right)\hat{\boldsymbol{e}}_z - \left(U\frac{\partial\bar{\alpha}_g}{\partial z}\right)\hat{\boldsymbol{e}}_\rho, \tag{4.18}$$

where

$$\frac{\partial\bar{\alpha}_g}{\partial z} = \frac{\rho\cos^2\bar{\alpha}_g}{(z + K\cos\bar{\alpha}_g)^2 + K\rho\sin\bar{\alpha}_g\cos^2\bar{\alpha}_g}, \tag{4.19}$$

$$\frac{\partial\bar{\alpha}_g}{\partial\rho} = -\frac{(z + K\cos\bar{\alpha}_g)\cos^2\bar{\alpha}_g}{(z + K\cos\bar{\alpha}_g)^2 + K\rho\sin\bar{\alpha}_g\cos^2\bar{\alpha}_g}, \tag{4.20}$$

$$U = \frac{3K\sin^2\bar{\alpha}_g\cos^2\bar{\alpha}_g + K\sin^4\bar{\alpha}_g + \rho\sin\bar{\alpha}_g}{2(\rho + K\sin^3\bar{\alpha}_g)^2}, \tag{4.21}$$

$$V = -\frac{K\sin^3\bar{\alpha}_g\cos\bar{\alpha}_g}{2\rho(\rho + K\sin^3\bar{\alpha}_g)^2}. \tag{4.22}$$

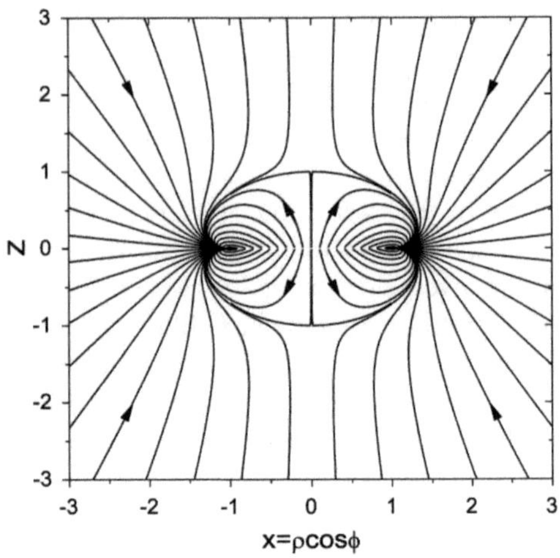

Fig. 4.1 Field-line distribution of $\boldsymbol{B}_g$ in the $x-z$ plane when $K=-1$. The arrows indicate the direction of $\boldsymbol{B}_g$

In addition, from the distribution of $\boldsymbol{A}_g$ at the large $\boldsymbol{R}$ limit, we find that a Dirac string exists along the positive z-axis.

Because the real antiferromagnetic ground state of the original system cannot be described under the present mean-field approximation, we focus on the ferromagnetic interaction case where $K<0$. Without loss of generality, we take $K=-1$. Then, the field line distribution of $\boldsymbol{B}_g$ (see Fig. 4.1) shows that the source and sink of $\boldsymbol{B}_g$ are located at the disk-shaped level-crossing region and its boundary, respectively. Specifically, one part of $\boldsymbol{B}_g$ originates from the level-crossing region and ends at its boundary, while the other part comes from infinity and also ends at its boundary. From (4.7) and the condition that $\gamma_g=2\pi$ or 0, with the evolution loop L being azimuthally symmetric, we find that the interface between these two parts, which we call the flux-free surface, is determined by $\rho=\sqrt{1-|z|}(1+\sqrt{|z|})$.

At the large-$\boldsymbol{R}$ limit, the interaction between the spins can be ignored. Then, the distribution of $\boldsymbol{B}_g$ is the field distribution of the point-like monopole with the elementary magnetic charge $(-1/2)$. This asymptotic distribution indicates that the source and sink of $\boldsymbol{B}_g$ as a whole can be interpreted as a disk-shaped monopole located at the region of the quantum phase transitions. Actually, similar deformed monopoles were reported in [9]. From the distribution of $\boldsymbol{B}_g$ near the level-crossing region, we find that the surface-density of the magnetic charge on this region is

$$\sigma_g=\frac{1}{4\pi(1-\rho^2)^{3/2}}, \tag{4.23}$$

where $\rho<1$. When $\rho\to 1$, (4.23) gives $\sigma_g\sim(1-\rho)^{-\Lambda}$ with the critical exponent $\Lambda=3/2$. The surface-density σ_g, together with the total magnetic charge of the monopole, determines $\boldsymbol{B}_g$ completely. Therefore, we can say that σ_g is the central quantity in the differential formulation for γ_g and that Λ is the corresponding central critical exponent. Since the Berry phase can provide the key ingredients of the criticality in principle [10], it is promising—albeit challenging—to investigate the relation between Λ and other critical exponents.

Because the ground state is twofold degenerate on the level-crossing region, any evolution loop L on this region corresponds to two different values of γ_g. It is straightforward to show that their difference $\Delta\gamma_g = 4\pi Q_g^L$ (modulo 2π), where Q_g^L is the magnetic charge enclosed by L. This finding indicates that the magnetic charge of the monopole originates from the phase difference $\Delta\gamma_g$. From (4.23), we find that the charge on the level-crossing region is divergent. Combining this divergence with the distribution of $\boldsymbol{B}_g$ at the large $\boldsymbol{R}$ limit, we find that the negative charge on the boundary of this region must also be divergent.

From the previous results, we can easily recognize that both the disk-shaped monopole and the exotic distribution of $\boldsymbol{B}_g$ are the natural consequences of the physical properties of the system. These properties include the following: (i) the asymptotic behavior of the system at the large $\boldsymbol{R}$ limit; (ii) the degeneracy of the ground state; and (iii) the criticality that leads to the divergent γ_g. Because similar properties exist in many interacting spin systems, the analogues of the disk-shaped monopole and its exotic magnetic-like field also exist generally in these systems [1].

On the other hand, the structure of the monopole and the distribution of $\boldsymbol{B}_g$ naturally reflect the properties of the system. In particular, the shape of the flux-free surface reflects the critical property of the system. Approximately speaking, both the external field $\boldsymbol{R}$ and the interaction between the spins affect γ_g, and the flux-free surface is located at the region where their effects on γ_g cancel each other out. In essence, the criticality of the system is exactly the consequence of the competition between $\boldsymbol{R}$ and the interaction. Therefore, the shape of the flux-free surface can reflect the critical property of the system, providing the possibility to measure the criticality by the Berry phase without having the system undergo the quantum phase transition.

4.2 Fractional Virtual Magnetic Monopole

We consider the following two-mode model, namely, the atom-diatomic molecule conversion system, as an example to demonstrate our theory [11]. The following deduction, in principle, can be extended to the case of multiple modes and multiatomic molecule formation. The energy of the system reads

$$\mathscr{H} = \frac{R\cos\theta}{2}\left(\hat{\psi}_1^\dagger\hat{\psi}_1 - \hat{\psi}_2^\dagger\hat{\psi}_2\right) + \sqrt{\frac{3}{8}}\frac{R\sin\theta}{2}\left(e^{i\varphi}\hat{\psi}_1^\dagger\hat{\psi}_1^\dagger\hat{\psi}_2 + h.c.\right), \quad (4.24)$$

where $h.c.$ denotes the Hermitian conjugate of the term in brackets and $\hat{\psi} = (\hat{\psi}_1, \hat{\psi}_2)$ and $\hat{\psi}^\dagger = (\hat{\psi}_1^\dagger, \hat{\psi}_2^\dagger)$ are the annihilation and creation operators for the atom and molecule, respectively. These expressions obey the commutation relations $[\hat{\psi}_i, \hat{\psi}_j^\dagger] = \delta_{ij}$ for bosons. Here, $\boldsymbol{R} = (R\sin\theta\cos\varphi, R\sin\theta\sin\varphi, R\cos\theta)$ is a vector in three-dimensional (3D) parameter space. The terms $\hat{\psi}_1^\dagger\hat{\psi}_1^\dagger\hat{\psi}_2 + h.c.$ describe the coupling between atom pairs and diatomic molecules, which brings a new gauge structure to the system. With these terms, the system is invariant under the transformation

$$U(\eta) = \mathrm{e}^{i\Theta(\eta)}, \quad \Theta(\eta) = \begin{pmatrix} \eta & 0 \\ 0 & 2\eta \end{pmatrix}. \tag{4.25}$$

Under the mean-field limit, i.e., replacing $\hat{\psi}$ and $\hat{\psi}^\dagger$ by complex numbers ψ and ψ^* that correspond to the coherent states of these operators, we rewrite $\mathscr{H}(\psi, \psi^*; \boldsymbol{R}) = \sum_{i,j} \psi_i^* T_{ij}(\psi, \psi^*; \boldsymbol{R})\psi_j$, where the matrix elements $T_{11} = -T_{22} = R\cos\theta/2$ and $T_{12} = T_{21}^* = \sqrt{3/8}(R\sin\theta/2)\mathrm{e}^{-i\varphi}\psi_1^*$, and obtain the nonlinear Schrödinger equation as follows ($\hbar = 1$):

$$i\frac{d}{dt}\begin{pmatrix} \psi_1 \\ \psi_2 \end{pmatrix} = H(\psi, \psi^*; \boldsymbol{R})\begin{pmatrix} \psi_1 \\ \psi_2 \end{pmatrix}, \tag{4.26}$$

with

$$H(\psi, \psi^*; \boldsymbol{R}) = \begin{pmatrix} \frac{R\cos\theta}{2} & \sqrt{\frac{3}{8}}\mathrm{e}^{i\varphi}R\sin\theta\,\psi_1^* \\ \sqrt{\frac{3}{8}}\mathrm{e}^{-i\varphi}R\sin\theta\,\frac{\psi_1}{2} & -\frac{R\cos\theta}{2} \end{pmatrix}. \tag{4.27}$$

The projective Hilbert space is spanned by the vector $\boldsymbol{n}_a = (2\sqrt{2}\mathrm{Re}(\psi_1^{*2}\psi_2), 2\sqrt{2}\mathrm{Im}(\psi_1^{*2}\psi_2), |\psi_1|^2 - 2|\psi_2|^2)$. Clearly, every point in this space corresponds to a class of quantum states, among which the states are only different in co-diagonal total phases (see (4.25)). With the normalization condition $|\psi_1|^2 + 2|\psi_2|^2 = 1$, one can find that the projection space is a teardrop-shaped surface [12].

Note that even though the matrix in (4.27) is not conjugate symmetric, the original system represented by (4.24) is Hermitian, and the total system energy is bound. The analogous nonlinear Schrödinger equations have been widely applied in the Feshbach molecular formation [6, 13].

The eigenstates of the above system $\bar{\phi}(\boldsymbol{R}) = (\bar{\phi}_1, \bar{\phi}_2)$ satisfy the following eigenequations:

$$H(\bar{\phi}, \bar{\phi}^*; \boldsymbol{R})\begin{pmatrix} \bar{\phi}_1 \\ \bar{\phi}_2 \end{pmatrix} = \begin{pmatrix} \mu & 0 \\ 0 & 2\mu \end{pmatrix}\begin{pmatrix} \bar{\phi}_1 \\ \bar{\phi}_2 \end{pmatrix}. \tag{4.28}$$

The above eigenequations define the eigenfunction $\bar{\phi}$ and the eigenvalue (or chemical potential) μ that are functions of the adiabatic parameter $\boldsymbol{R}$. The eigenequations are solved, and the eigenfunctions are obtained as follows:

$$\bar{\phi}_2^{\pm} = \frac{-\cos\theta \pm 1}{\sqrt{6}\sin\theta}, \quad \bar{\phi}_1^{\pm} = \mathrm{e}^{i\varphi/2}\sqrt{1 - 2|\bar{\phi}_2^{\pm}|^2}, \tag{4.29}$$

with the eigenvalue (or chemical potential for an atom) $\mu_{\pm} = \frac{R}{4}(\cos\theta \pm 1)$.

Now, let us suppose that the parameter vector $\boldsymbol{R}(t)$ varies slowly in a time interval $t \in [0, T]$. For simplicity, we fix R and θ, and we change φ slowly from 0 to 2π, forming a loop in the parameter space, i.e., $\boldsymbol{R}(0) = \boldsymbol{R}(T)$. The sweep rate $\alpha = 2\pi/T$ measures how slowly the system evolves, and $\alpha \to 0$ indicates the adiabatic limit. Initially the system populates in an eigenstate, i.e., $\psi(0) = \bar{\phi}(\boldsymbol{R}(0))$, and remains close to the eigenstate during an adiabatic process, as ensured by the adiabatic theory [6, 14]. That is, $||\psi(t) - \bar{\phi}(\boldsymbol{R}(t))|| \equiv 1 - |\langle\psi(t)|\bar{\phi}(\boldsymbol{R}(t))\rangle|^2 \sim \alpha^2$. This result indicates that $|\langle\psi(T)|\psi(0)\rangle|^2 = 1$ in the adiabatic limit. Moreover, a total phase is acquired over the course of the cycle. As we show below, the total phase (for the atomic component) $\arg(\langle\psi(T)|\psi(0)\rangle)$ can be decomposed into two parts in the adiabatic limit, i.e., $\lim_{\alpha\to 0}\arg(\langle\psi(T)|\psi(0)\rangle) = -(\gamma_d + \gamma_g)$. The former is dynamical and can be expressed as the time integral of the chemical potential for an atom; the latter is of geometric property, and its explicit expression will be derived analytically and shown to be dramatically different from the usual Berry phase formula.

We find that (4.26) and its conjugate construct a canonical structure of classical dynamics with the energy $\mathcal{H}(\psi, \psi^*; \boldsymbol{R})$ as a classical Hamiltonian and $(\psi, i\psi^*)$ as a canonical variable pair. The gauge symmetry of $\mathcal{H}$ given by (4.25) implies that the total atom number is conserved $|\psi_1|^2 + 2|\psi_2|^2 = 1$ and that the dynamics of the overall phase can be separated from the rest of the degrees of freedom [5]. For simplicity and without loss of generality, we denote $\lambda = \arg\psi_1$ and set the total phase as λ for the atomic component and 2λ for the molecular component. The other variables form a closed set of Hamiltonian dynamics with the canonical pair defined as $q = -\arg\psi_2 + 2\arg\psi_1$ and $p = |\psi_2|^2$. From (4.26) and its complex conjugate, we obtain

$$\frac{d\lambda}{dt} = p\frac{dq}{dt} - \mathcal{H}(p, q) - \Lambda(p, q), \tag{4.30}$$

$$\frac{dp}{dt} = -\frac{\partial\mathcal{H}}{\partial q}, \quad \frac{dq}{dt} = \frac{\partial\mathcal{H}}{\partial p}, \tag{4.31}$$

in which $\mathcal{H}(p, q) = R\cos\theta(1 - 3p)/2 + \sqrt{3/8}R\sin\theta(1 - 2p)\sqrt{p}\cos(q - \varphi)$, $\Lambda(p, q) = \mathrm{Re}[\sum_{i,j,k}\sqrt{p_i}(p_j\partial\tilde{T}_{ik}/\partial p_j - i\partial\tilde{T}_{ik}/\partial q_j)\sqrt{p_k}]$ and $\tilde{T}_{ik} = \mathrm{e}^{i(\arg\psi_k - \arg\psi_i)} T_{ik}$. After a series of calculations, we obtain $\Lambda(p, q) = \sqrt{3/8}(R\sin\theta/2)(1 - 2p)\sqrt{p}\cos(q - \varphi)$. The fixed point of the above Hamiltonian can be derived by setting the right-hand functions equal to zero in (4.31). We readily obtain $\bar{p} = |\bar{\phi}_2^{\pm}|^2$ and $\bar{q} = \varphi$. These fixed points are local energy minima of the system and therefore correspond to the eigenstates defined by (4.28).

For a linear quantum case, both matrices $\{H_{ij}\}$ and $\{T_{ij}\}$ are functions of the parameter $\boldsymbol{R}$ only; thus, the last term in (4.30) vanishes, i.e., $\Lambda(p, q) = 0$. The second term on the right-hand side of (4.30) is the energy, whose time integral gives the so-called dynamical phase. The time integral of the first term is the Aharonov-Anandan phase for a cyclic quantum evolution [15]. The above observation is readily extended to the adiabatic evolution of a quantum eigenstate because the adiabatic theorem of quantum mechanics dictates that an initial nondegenerate eigenstate remains an instantaneous eigenstate and that the evolution is cyclic when the parameters move slowly in a circuit. In this case, the second term is the eigenenergy, and the first term is just the Berry connection, i.e., $i\langle\bar{\phi}(\boldsymbol{R})|\nabla|\bar{\phi}(\boldsymbol{R})\rangle$. The Berry phase equals the circuit integral of the Berry connection.

However, for our atom-molecule system, the contribution of the last term in (4.30) should be taken into account. Note that the chemical potential is typically not identical to the energy, while the dynamical phase represents the time integral of the chemical potential, and we need to evaluate the following quantity in the adiabatic limit:

$$\Xi(p,q;\boldsymbol{R}) = \mathscr{H}(p,q) + \Lambda(p,q) - \mu(\boldsymbol{R}). \tag{4.32}$$

We set $p = \bar{p}(\boldsymbol{R}) + \delta p$ and $q = \bar{q}(\boldsymbol{R}) + \delta q$. Here, $\bar{p}(\boldsymbol{R})$ and $\bar{q}(\boldsymbol{R})$ are the fixed points corresponding to the eigenstates defined by (4.28). The vector $(\delta p, \delta q)$ represents the corrections to the adiabatic eigenstates in the order of α [16]. As will be shown, $(\delta p, \delta q)$ contain secular terms in addition to the rapid oscillations. These secular terms are accumulated in the nonlinear adiabatic evolution and contribute to the geometric phase.

We expand the quantity $\Xi(p,q;\boldsymbol{R})$ around the fixed point:

$$\Xi(p,q;\boldsymbol{R}) = \sqrt{\frac{3}{8}}\frac{R\sin\theta}{2}\frac{1-6\bar{p}}{2\sqrt{\bar{p}}}\delta p + O(\delta p^2, \delta q^2). \tag{4.33}$$

Here, we have used the relations $\mathscr{H}(\bar{p},\bar{q}) + \Lambda(\bar{p},\bar{q}) = \mu(\boldsymbol{R})$, $\partial\mathscr{H}(p,q)/\partial p|_{(\bar{p},\bar{q})} = \partial\mathscr{H}(p,q)/\partial q|_{(\bar{p},\bar{q})} = 0$ and $\partial\Lambda/\partial q|_{(\bar{p},\bar{q})} = 0$. On the other hand, $(\delta p, \delta q)$ can be evaluated from the following Hamiltonian equations:

$$\frac{dq}{dt} = \left.\frac{\partial^2\mathscr{H}}{\partial p\partial p}\right|_{(\bar{p},\bar{q})}\delta p + \left.\frac{\partial^2\mathscr{H}}{\partial p\partial q}\right|_{(\bar{p},\bar{q})}\delta q + O(\delta q^2, \delta p^2), \tag{4.34}$$

$$\frac{dp}{dt} = -\left.\frac{\partial^2\mathscr{H}}{\partial q\partial p}\right|_{(\bar{p},\bar{q})}\delta p - \left.\frac{\partial^2\mathscr{H}}{\partial q\partial q}\right|_{(\bar{p},\bar{q})}\delta q + O(\delta q^2, \delta p^2). \tag{4.35}$$

By averaging the fast oscillations, one can omit the higher-order terms and retain the secular terms; then, $(\langle\delta p\rangle, \langle\delta q\rangle)^T = \Omega^{-1}((\partial\bar{p}/\partial\boldsymbol{R})(d\boldsymbol{R}/dt), (\partial\bar{q}/\partial\boldsymbol{R})(d\boldsymbol{R}/dt))^T$, with the matrix $\Omega = \frac{\sqrt{8}}{\sqrt{3}R\sin\theta}\begin{pmatrix} 0 & -2\bar{p}\sqrt{\bar{p}}/(1+6\bar{p}) \\ 1/[(1-2\bar{p})\sqrt{\bar{p}}] & 0 \end{pmatrix}$. We then have $\delta p = -(\sqrt{8}/(\sqrt{3}R\sin\theta))(2\bar{p}\sqrt{\bar{p}}/(1+6\bar{p}))(d\boldsymbol{R}/dt)$. Therefore, by substituting the result into (4.33), from (4.30), we obtain the total phase acquired by the eigenstate in the adiabatic limit, $\lambda = -\gamma_d + \gamma_g$ with $\gamma_d = \int\mu dt$, and the geometric phase,

$$\gamma_g = \oint\bar{p}d\varphi + \oint\frac{(1-6\bar{p})\bar{p}}{1+6\bar{p}}d\varphi \tag{4.36}$$

$$= \frac{1}{6}\oint(1\mp\cos\theta)d\varphi. \tag{4.37}$$

In contrast to the previous works [17], the adiabatic geometric phase in the atom-molecule system is dramatically modified. The first term on the right-hand side of (4.36) is the typical expression of the Berry phase that can be rewritten as the circuit integral of the Berry connection $i\langle\bar{\phi}(\boldsymbol{R})|\nabla|\bar{\phi}(\boldsymbol{R})\rangle$. The second novel term indicates that the higher-order correction to the adiabatic approximate solution (which is negligible in the linear case) can be accumulated in the nonlinear adiabatic evolution with an infinite time duration in the adiabatic limit and contributes a finite phase with a geometric nature.

The above theoretical formulation on the Berry phase has been verified numerically by directly integrating the Schrödinger equation along the circle path in parameter space with both R and θ fixed and with φ varying with a very small sweeping rate α, i.e., $\varphi = \alpha t$. In Fig. 4.2, we show the numerical results and compare them with theoretical predictions for the adiabatic geometric phase of the eigenstate of μ_+, in which the sweeping rate is $\alpha = 0.0001$. The inset figure shows the convergence of the adiabatic geometric phase with the sweeping rates.

For linear systems, the Berry phase has been interpreted as the flux of a magnetic field of a quantized monopole through the surface enclosed by a loop in parameter space. This interpretation is demonstrated by the spin-half system, i.e., $H = \boldsymbol{R} \cdot \boldsymbol{\sigma}/2$, where $\boldsymbol{\sigma}$ denotes Pauli matrices and $\boldsymbol{R} = (R \sin\theta \cos\varphi, R \sin\theta \sin\varphi, R\cos\theta)$ is a vector in the 3D parameter space. The eigenvalues of the system are $E = \pm R/2$, and their eigenstates are $|\psi_\pm\rangle = (\pm\sqrt{(1 \pm \cos\theta)/2}e^{-i\varphi/2}, \sqrt{(1 \mp \cos\theta)/2}e^{i\varphi/2})^T$. The Berry phase equals the circuit integral of the Berry connection $i\langle\psi(\boldsymbol{R})|\nabla|\psi(\boldsymbol{R})\rangle$ and can be interpreted as the flux of the magnetic field of a virtual monopole with charge $g_0 = 1/2$. In general, the degeneracies of the spectrum in parameter space

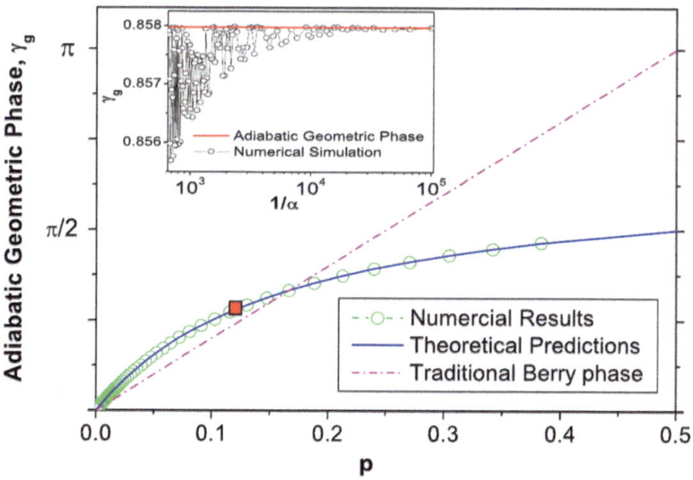

Fig. 4.2 Comparison between the numerical results and the theoretical predictions for the adiabatic geometric phase of the eigenstate for μ_+, in which the sweeping rate is $\alpha = 0.0001$. The insert figure shows the convergence of the geometric phase with the sweeping rate

are the singularities of the virtual magnetic field, and these degeneracies therefore play an important role in connection with the geometric phase. Each degeneracy can be interpreted as a charge distribution located at the contact point between energy surfaces. Because the eigenstates are smooth and single-valued outside the degeneracies, the total charge of the distribution, i.e., the monopole charge, is necessarily an integer multiple of the elementary charge $g_0 = 1/2$. Nonelementary monopoles with integer multiples of g_0 have been found in the cases of light propagation and condensed matter physics [18, 19]. The mechanism for the production of monopole charges that are larger than the elementary g_0 is due to the constraints that act on the system [20].

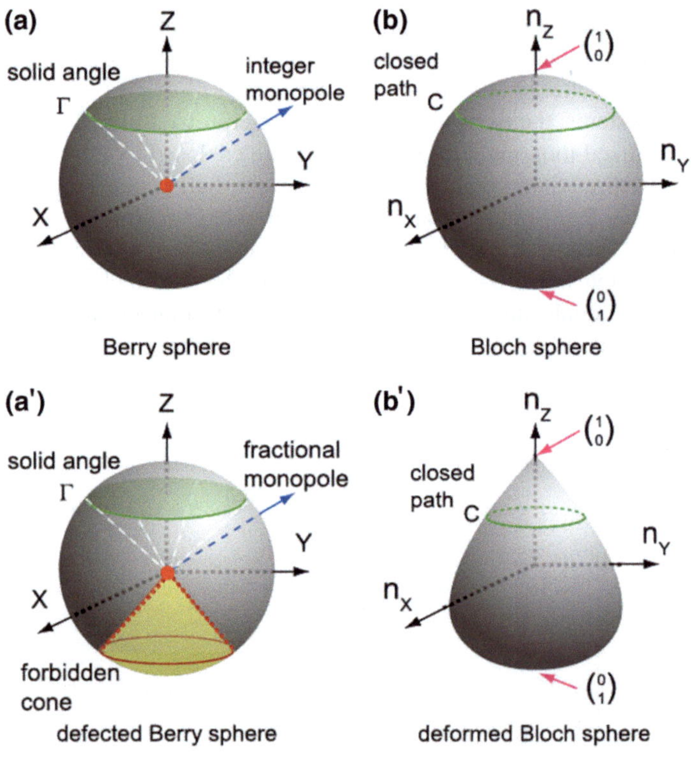

Fig. 4.3 Parameter space (**a**) and Bloch sphere (**b**) for a spin-half particle in a magnetic field. **a**′ and **b**′ are the parameter space and Bloch sphere for the atom-molecule conversion system, respectively. The parameters change adiabatically along a closed path, shown as the green circles in the parameter spaces or the Berry spheres. Accordingly, the eigenstate evolves and forms a closed path schematically plotted as green circles on the Bloch spheres. The gray cone in (**a**′) is the boundary for which $\theta = 2\pi/3$, inside which (corresponding to $\theta > 2\pi/3$) no eigenstate exists. See the text for details

For our nonlinear system, when the parameters $\boldsymbol{R} = (R, \theta, \varphi)$ are considered as the spherical coordinates of a vector in a 3D space, from (4.37), we obtain the vector potential

$$\boldsymbol{A} = \frac{1}{6}\frac{(1-\cos\theta)}{R\sin\theta}\hat{\boldsymbol{e}}_{\varphi}, \tag{4.38}$$

where $\hat{\boldsymbol{e}}_{\varphi}$ is the unit vector in the direction φ. For convenience, we consider only the case for μ_+. Hence, the Berry phase can be interpreted as the flux of a magnetic field of a virtual monopole through the surface enclosed by a closed path in parameter space (see Fig. 4.3a′), $\boldsymbol{B} = \nabla \times \boldsymbol{A} = g\boldsymbol{R}/R^3$. Here, the monopole charge $g = g_0/3$ is the fractional elementary charge of the quantized monopole. The fractional charge of the monopole in this system is due to the symmetry breaking of the parameter space by the boundary. From (4.29), we find that $|\bar{\psi}_2|$ monotonically increases with increasing θ. When $\theta = 2\pi/3$, $|\bar{\psi}_2| = 1/\sqrt{2}$ reaches its extreme value (since $|\bar{\psi}_1|^2 + 2|\bar{\psi}_2|^2 = 1$). This result implies that no eigenstate exists in the regime $\theta > 2\pi/3$, i.e., the Berry sphere of this system is a deformed sphere with a forbidden cone bounded by $\theta = 2\pi/3$. We illustrate this structure in Fig. 4.3a′. The strange string lies on the negative Z-axis, i.e., $\theta = \pi$.

4.3 Virtual Magnetic Monopole Chain

In this section, we discuss the monopoles in a bosonic atom–heteronuclear-molecule conversion system. This condition may be even more interesting than the atom–homonuclear-molecule system discussed in the proceeding section because the heteronuclear molecules can be either bosons or fermions; thus, quantum statistics can be expected to play an important role [21], and a large electric dipole moment will be induced with the potential to create a dipolar superfluid [22]. In the quantum theory, we calculate the virtual monopole field and its charge for the ground state. In particular, we explore the effects of the particle-number imbalance between two atomic species on the monopole field. We find that the monopole field in this system is not spherically symmetrical and that the charge of the monopole depends strongly on the particle-number imbalance [23].

We adopt a three-mode model to describe an atom–heteronuclear-molecule system. The basic assumption here is that the spatial wave functions for three modes are fixed so that one can associate each mode with an annihilation operator $\hat{a}_j$ of a particle in atomic modes $j = 1$ and 2 and in molecular mode $j = m$. Under this three-mode approximation, the second-quantized Hamiltonian of the system reads (with $\hbar = 1$ throughout) [24]

$$\hat{H} = \Delta\hat{N}_m + \sum_{i,j}\frac{\chi_{ij}}{N}\hat{N}_i\hat{N}_j + \left(\frac{\Omega e^{i\varphi}}{\sqrt{N}}\hat{a}_1^{\dagger}\hat{a}_2^{\dagger}\hat{a}_m + h.c.\right), \tag{4.39}$$

where $h.c.$ denotes the Hermitian conjugate of the term in brackets. $\hat{N}_j = \hat{a}_j^\dagger \hat{a}_j$. The detuning Δ represents the energy difference between the molecular and atomic levels, which can be tuned by an external field. The parameter $\Omega e^{i\varphi}$ specifies the atom-molecule coupling. The parameters χ_{ij} describe s-wave scattering, taking into account the intraspecies ($i = j$) and interspecies ($i \neq j$) with $\chi_{ij} = \chi_{ji}$. Note that the model (4.39) can be mapped to a trilinear Hamiltonian describing the nondegenerate parametric down-conversion in quantum optics [25]. In this analogy, the molecular mode plays the role of the pump photon, while the two atomic modes denote the signal and idler photons. The collisional terms correspond to the Kerr-type cubic nonlinearity that is present in the optical system with some nonlinear medium [26]. In experiments, the model (4.39) may apply to the ^{85}Rb-^{87}Rb system in which the ultracold heteronuclear Feshbach molecules were produced starting with a ^{87}Rb Bose-Einstein condensate (BEC) and a cold atomic gas of ^{85}Rb [27]. These ultracold heteronuclear molecules in low-lying vibrational states are of particular interest since they could be a permanent dipole moment due to the unequal distribution of electrons. The Hamiltonian (4.39) commutes with $\hat{N} = \hat{a}_1^\dagger \hat{a}_1 + \hat{a}_2^\dagger \hat{a}_2 + 2\hat{a}_m^\dagger \hat{a}_m$; thus, the total atomic number $N = \langle \hat{N} \rangle$ is a conserved quantity of the system. Indeed, there exists another conserved quantity, namely, $D = \langle \hat{a}_1^\dagger \hat{a}_1 \rangle - \langle \hat{a}_2^\dagger \hat{a}_2 \rangle$, which denotes the particle-number imbalance between two atomic species. Using these two conserved constants and neglecting the trivial constant terms that are proportional to D or N, the Hamiltonian (4.39) can be simplified as follows:

$$\hat{H} = Z\hat{N}_m + \frac{\chi}{N}\hat{N}_m^2 + \left(\frac{X + iY}{\sqrt{N}} \hat{a}_1^\dagger \hat{a}_2^\dagger \hat{a}_m + H.c. \right), \tag{4.40}$$

where $X + iY = \Omega e^{i\varphi}$, $Z = \Delta - (D + N)(\chi_{11} - \chi_{1m}) - 2N\chi_{12} + (D - N)(\chi_{22} - \chi_{2m})$, and $\chi = \chi_{11} + \chi_{22} + \chi_{mm} + 2(\chi_{12} - \chi_{1m} - \chi_{2m})$. For our atom-molecule system, the Bloch space is expanded by the following three angular momentum operators: $\hat{L}_X = 2\sqrt{2}\hat{a}_1^\dagger \hat{a}_2^\dagger \hat{a}_m + \hat{a}_m^\dagger \hat{a}_1 \hat{a}_2 / N^{3/2}$, $\hat{L}_Y = i2\sqrt{2}\hat{a}_1^\dagger \hat{a}_2^\dagger \hat{a}_m - \hat{a}_m^\dagger \hat{a}_1 \hat{a}_2 / N^{3/2}$, and $\hat{L}_Z = 2\hat{a}_m^\dagger \hat{a}_m - \hat{a}_1^\dagger \hat{a}_1 - \hat{a}_2^\dagger \hat{a}_2 / N$. These operators compose a generalized Bloch representation [28, 29], and the Hamiltonian in this representation (4.40) becomes $\hat{H} = (N/4)(Z + \frac{\chi}{2})\hat{L}_Z + (N\chi/16)\hat{L}_Z^2 + [N/(4\sqrt{2})][(X + iY)(\hat{L}_X - i\hat{L}_Y) + h.c.]$. The corresponding parameter space is spanned by the vector $\boldsymbol{R} = (X, Y, Z)$, which represents the influence of an external field.

In the following discussion, we focus only on the monopoles for the ground states. For convenience, we restrict ourselves to the states with even N and D, where $D = 0, 2, \ldots, N - 2$. From the Hamiltonian (4.40), one can compute the ground-state MB curvature (i.e., the virtual magnetic field) by using the formula [30]

$$\boldsymbol{B}_N(\boldsymbol{R}) = \mathrm{Im} \sum_{m \neq 0} \frac{\langle 0 | \nabla_{\boldsymbol{R}} \hat{H} | m \rangle \times \langle m | \nabla_{\boldsymbol{R}} \hat{H} | 0 \rangle}{(E_m - E_0)^2}, \tag{4.41}$$

where $|0\rangle$ denotes the quantum ground state. E_m specifies the eigenenergy of the eigenstate $|m\rangle$ and satisfies the eigenequation $\hat{H}|m\rangle = E_m|m\rangle$. The energy denominator in (4.41) implies that the virtual magnetic field typically diverges at the degenerate point where the energy levels cross and may be maximized at avoided level crossings. These level structures are reflected in the geometry of the Hilbert space of the system, which can be captured by the virtual magnetic field of the ground-state monopole. When particle interaction is absent, i.e., $\chi = 0$, only a point-like magnetic monopole exists in our system, which is located at the origin $\boldsymbol{R} = 0$ in the parameter space. We have numerically computed the virtual magnetic field of the ground-state monopole from (4.40) and (4.41) by using the exact diagonalization method in double-precision arithmetic in the Fock-state representation. The results are shown in Fig. 4.4a–c. We find that the virtual magnetic field is symmetric with respect to the Z-axis in our atom-molecule system, which is different from the isotropic field generated by a standard point-like monopole. However, the symmetry of the virtual field approaches spherical symmetry as the atom-number imbalance parameter D increases. When $D = N - 2$, a spherically symmetric field is recovered. The similarity between the upper panels and the lower panels of Fig. 4.4 reveals that the properties of the virtual field are determined mainly by the structure of the level gap between the ground state and the first excited state [23].

In the presence of the particle interaction, i.e., $\chi \neq 0$, the ground state of the system exhibits a monopole chain that includes $(N - D)/2$ point-like monopoles on the Z axis, which are located at the points $(X = Y = 0, Z/\chi = -n/N)$ with $n = N - D - 1, N - D - 3, \ldots, 1$. When the particle-number imbalance between two atomic species increases, the number of monopoles decreases. When $D = N - 2$, there is

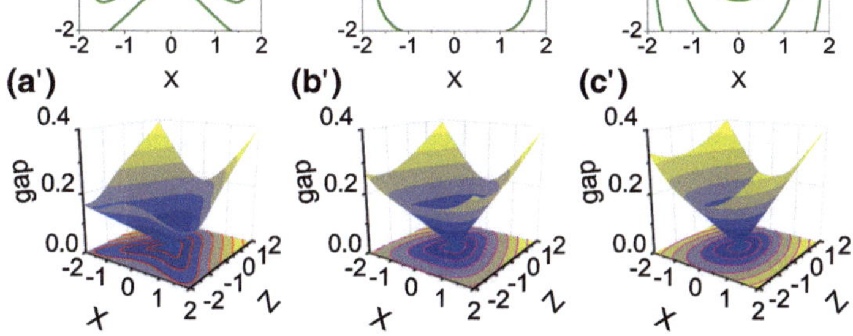

Fig. 4.4 (Color online) Magnitude of the virtual field for the ground state (upper panels) and the level gap between the ground state and the first exited state (lower panels) with $N = 10$ and $\chi = Y = 0$. **a** and **a**$'$: $D = 0$; **b** and **b**$'$: $D = 2$; **c** and **c**$'$: $D = 4$. In the upper panels, the contours denote the rescaled quantity $\log_{10} |\boldsymbol{B}_N/N|$, and the values are -1.5, -1.2, -0.9, -0.6, -0.1, and 0.5 from the outside to the core

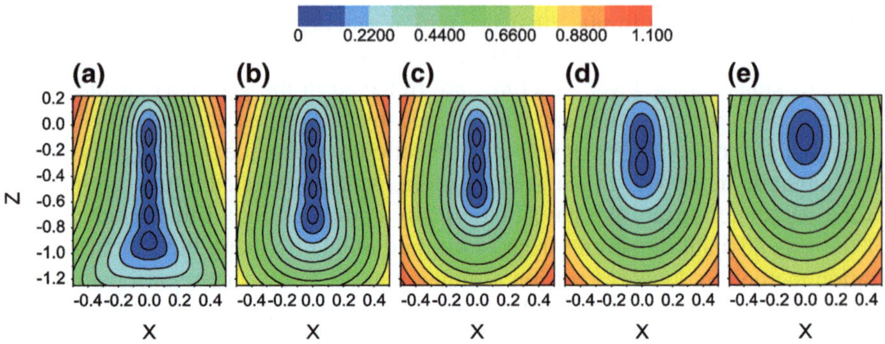

Fig. 4.5 (Color online) Level gap between the ground state and the first excited state with $N = 10$ and $Y = 0$. From **a** to **e**, $D = 0, 2, 4, 6$, and 8. The interaction parameter $\chi = 1$ has been used

only one monopole located at the point ($X = Y = 0, Z/\chi = -1/N$). The level gap between the ground state and the first excited state for this case is demonstrated in Fig. 4.5.

We now numerically calculate the monopole charge. According to Gauss's law, the charge g_N of the monopole can be defined by

$$\oint_S \boldsymbol{B}_N \cdot d\boldsymbol{S} = 4\pi g_N, \tag{4.42}$$

where $d\boldsymbol{S}$ is the area element and S denotes any surface boundary enclosing all monopole points. To numerically calculate the monopole charge from (4.42), we first compute the monopole field from (4.40) and (4.41) with the exact diagonalization method, and then, we choose a sphere as the closed surface enclosing all monopoles. With increasing radius of the sphere, the numerical tests demonstrate satisfactory convergence and accuracy of the results for g_N. The results of the monopole charge for $\chi = 0$ with different total atom numbers are demonstrated in Fig. 4.6. Note that the monopole charge in our system is not equal to the elementary charge $g_0 = 1/2$. In the special case $D = 0$, the charge $g_N = -N/4 = -Ng_0/2$. In the general case $D \neq 0$, the charge of the monopole decreases smoothly from $-Ng_0/2$ to $-g_0$ as the atom-number imbalance between the two atomic species increases. The general formula for the monopole charge of the ground state in our system is given by [23]

$$g_N = -\frac{N-D}{4} = -\frac{N-D}{2} g_0, \tag{4.43}$$

which is determined by the property of $((N-D)/2+1)$-fold degeneracy of the ground states; the degenerate states are $|(N+D-2m)/2, (N-D-2m)/2, m\rangle$ with $m = 0, 1, \ldots, (N-D)/2$. This result implies that one can create a magnetic monopole with nonelementary monopole charge in an atom–heteronuclear-molecule system. Note that the total charge of the monopole for the ground state in the second-quantized

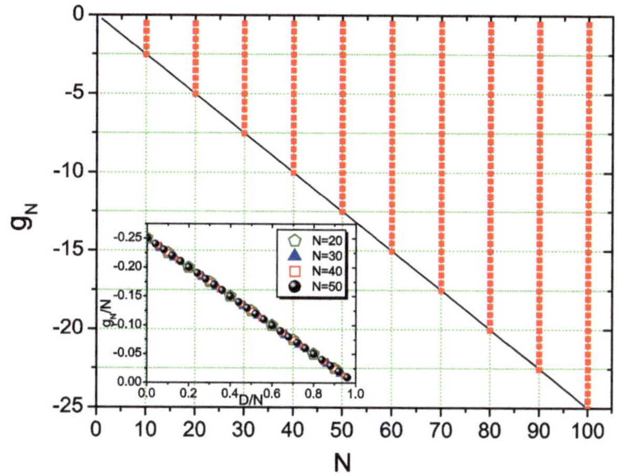

Fig. 4.6 (Color online) Monopole charge of the ground state in the model (4.40) as a function of the total particle number N, with $\chi = 0$. For each N, the points shown from bottom to top correspond to the cases $D = 0, 2, \ldots$, and $N - 2$. Inset shows the rescaled charge as a function of the rescaled population imbalance between the two atomic species

model is $Q = 4\pi g_N = -(N - D)\pi$ and that the Chern number is $-(N - D)/2$. We find that the Chern number is also an integer in our system, which is similar to the result derived for an interacting boson system [9].

When $\chi \neq 0$, we choose a closed surface that encloses all $(N - D)/2$ degenerate points, and the monopole charge of the ground state for $\chi = 1$ is obtained, as shown in Fig. 4.7. For $D = 0$, the number of the degenerate points is $N/2$, and the charge is $-N/4$. For $D \neq 0$, the degenerate points reduce to $(N - D)/2$, and the charge becomes $-(N - D)/4$. This result implies that the result (4.43) obtained for $\chi = 0$ is also appropriate for the case $\chi \neq 0$. Clearly, the total charge of the monopoles for the ground state is also $-(N - D)\pi$, and the Chern number is also $-(N - D)/2$. In fact, the quantum ground state is doubly degenerate at $(N - D)/2$ points in this case. At each degenerate point, the two degenerate eigenstates are $|(N + D - n + 1)/2, (N - D - n + 1)/2, (n - 1)/2\rangle$ and $|(N + D - n - 1)/2, (N - D - n - 1)/2, (n + 1)/2\rangle$.

In the mean-field limit, i.e., $N \to \infty$, the atom-molecule system becomes classical [31] and can be accurately described by the following semiclassical Hamiltonian [24, 32]:

$$H_c = \lim_{N\to\infty} \frac{\langle \hat{H} \rangle}{N} = Z|a_m|^2 + \chi|a_m|^4 + [(X + iY)a_1^* a_2^* a_m + c.c.], \tag{4.44}$$

where $c.c.$ denotes the complex conjugate of the term in brackets. Here, a_j are complex amplitudes for the system in the three quantum modes, and the normalized condition is given by $|a_1|^2 + |a_2|^2 + 2|a_m| = 1$.

Note that the mean-field model (4.44) does not admit the $U(1)$ gauge transformation, which is an especially interesting point about our atom-molecule system. In fact, this model is invariant under the gauge transformation as follows:

$$|\psi\rangle \to U(\phi_1, \phi_2)|\psi\rangle = e^{i\Theta(\phi_1, \phi_2)}|\psi\rangle, \tag{4.45}$$

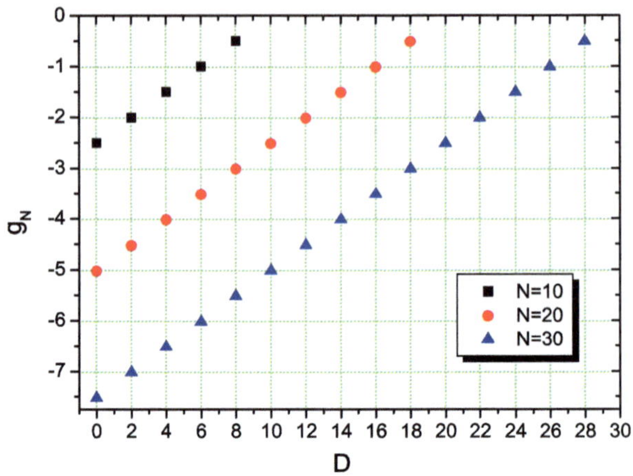

Fig. 4.7 (Color online) Monopole charge of the ground state in the model (4.40) as a function of the population imbalance parameter D, with $\chi = 1$

where

$$\Theta(\phi_1, \phi_2) = \begin{pmatrix} \phi_1 & 0 & 0 \\ 0 & \phi_2 & 0 \\ 0 & 0 & \phi_1 + \phi_2 \end{pmatrix}. \tag{4.46}$$

This type of gauge transformation includes two nonidentical phase parameters (i.e., ϕ_1 and ϕ_2), which is different from the so-called skewed $U(1)$ gauge transformation introduced in [11, 33], where the transformation depends on only one parameter. However, we observe that the nonlinear model (4.44) has a classical Hamiltonian structure (i.e., we can introduce three pairs of conjugate variables through $p_j = \sqrt{i}a_j^*$ and $q_j = \sqrt{i}a_j$). Following [33, 34], we can define for our system the Berry connection $\boldsymbol{A}$ and determine its general properties by making a gauge transformation as follows:

$$\begin{aligned} \boldsymbol{A} = i\langle\psi|\bar{\nabla}_{\boldsymbol{R}}|\psi\rangle \rightarrow \boldsymbol{A}' &= i\langle\psi'|\bar{\nabla}_{\boldsymbol{R}}|\psi'\rangle \\ &= \boldsymbol{A} - (|a_1|^2 + |a_2|^2 + 2|a_m|^2)\nabla_{\boldsymbol{R}}\phi_1 \\ &\quad + \frac{1}{2}[1 - (|a_1|^2 - |a_2|^2)]\nabla_{\boldsymbol{R}}(\phi_1 - \phi_2), \end{aligned} \tag{4.47}$$

where $|\psi'\rangle = U(\phi_1, \phi_2)|\psi\rangle$ and the overbar indicates an average over all initial angles with the same actions. It is found that the second term and the last term in (4.47) are trivial total derivatives due to the conservation of the total particle number and the constant particle-number imbalance between two atomic species (i.e., $d = |a_1|^2 - |a_2|^2$). This result implies that the Berry connection defined above is gauge invariant under the transformation $U(\phi_1, \phi_2)$. For an instantaneous eigenstate, the unnecessary averaging operation can be safely neglected because the population probabilities of different eigenstates are just the classical actions, which are found to be the adiabatic constants in an adiabatic evolution [5].

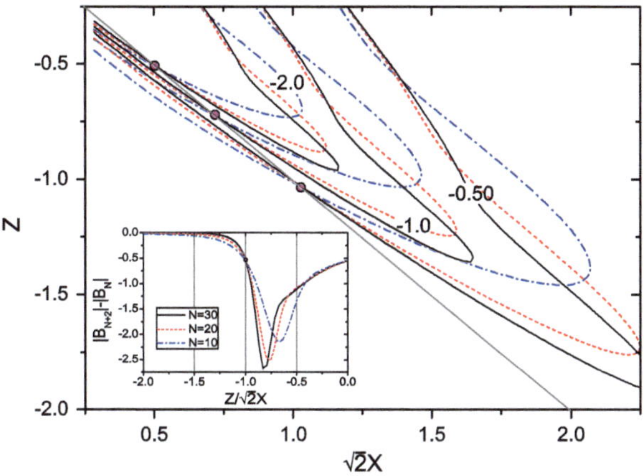

Fig. 4.8 (Color online) Contour lines of $(|\boldsymbol{B}_{N+2}| - |\boldsymbol{B}_N|)$ for the ground state at $Y = 0$, with $\chi = D = 0$. Each set of contour lines with the same value includes the data with $N = 10$ (blue dashed-dotted line), $N = 20$ (red dashed line), and $N = 30$ (black solid line). Solid circles indicate the intersection points. Inset shows the dependence of $(|\boldsymbol{B}_{N+2}| - |\boldsymbol{B}_N|)$ on the parameter $Z/\sqrt{2}X$ for different N. The gray straight line (with slope -1) is plotted to guide the eye

From the definition (4.47), one can calculate the mean-field curvature $\boldsymbol{B}$ and the monopole charge g for the ground state through $\boldsymbol{B} = \nabla_{\boldsymbol{R}} \times \boldsymbol{A}$ and its closed surface integral, respectively. Actually, the mean-field and quantum virtual fields satisfy the relation $\lim_{N\to\infty}(\boldsymbol{B}_N/N - \boldsymbol{B}) = 0$, which can be proved from the fact that H_c is the semiclassical limit of the second-quantized model $\hat{H}$ (for details, see [9, 33]) and by following Berry's argument about the semiclassical connection between the Berry phase and the Hannay angle [35, 36]. Our numerical simulations have shown that this relation indeed holds everywhere in the parameter space except at the degenerate points. For the monopole charge, we have

$$\lim_{N\to\infty}\left(\frac{g_N}{N} - g\right) = 0, \tag{4.48}$$

with

$$g = \lim_{N\to\infty}\frac{g_N}{N} = -\frac{1}{4}(1-d) = -\frac{1}{2}(1-d)g_0, \tag{4.49}$$

where the population-imbalance parameter $d = \lim_{N\to\infty} D/N$. Equation (4.49) implies that the ground-state monopole in our system in the mean-field limit can carry an arbitrary charge ranging from zero to $-(1-d)/4$ (i.e., the discrete points in Figs. 4.6 and 4.7 are connected to form continuous lines).

We now present selected discussion points. When $D = 0$, our system reduces to the homonuclear molecule system. In this case, the system exhibits a quantum phase transition from a mixture phase to a pure molecule phase at the critical point $Z_c = -\sqrt{2(X^2+Y^2)}$ [37]. To show the connection of the virtual field to the quantum phase transition, we plot the contour lines of the magnitude ($|\boldsymbol{B}_{N+2}| - |\boldsymbol{B}_N|$) for the ground state in Fig. 4.8. We find that the contours with the same values for different particle numbers cross at the phase transition point. This result is independent of the system size, which implies that even though a quantum phase transition is only rigorously defined in the thermodynamic limit $N \to \infty$, the virtual field of the monopole exactly marks the changes in the ground states of the system for a finite particle-number case [38]. However, in the case of $D \neq 0$, no quantum phase transition occurs even in the semiclassical limit [24].

In fact, the degeneracies of the spectrum in the parameter space are the singularities of the virtual field, and therefore, the monopoles play an important role in connection with the geometric phase. Each degeneracy can be interpreted as a charge distribution located at the contact point between energy surfaces. Because the eigenstates are smooth and single valued outside the degeneracies, the monopole charge is necessarily an integer multiple of the elementary charge g_0. In the generic case of a diabolical contact [30], the monopole charges are precisely $\pm g_0$. However, higher integer multiples of g_0 may occur [39]. For instance, for light propagating through a twisted anisotropic dielectric medium, experimental situations arise [18] in which the monopole charges are $\pm 2g_0$. Our present discussion provides a perfect example for higher integer multiples of g_0 in ultracold atom-molecule systems. We emphasize that even though our results are obtained with a specific three-level boson model, the results are expected to hold in a general interacting atom–heteronuclear-molecule boson system where many heteronuclear molecules in high vibrational states are included. The reason is that our system is the simplest atom–heteronuclear-molecule system; we consider only the zero-temperature homogeneous case so that all the bosons are condensed into zero center-of-mass momentum states. In particular, the monopole as the degeneracy point in the Brillouin zone is found to play a pivotal role in the anomalous Hall effect [40]. This fact implies that it may be possible to observe our predicted virtual monopole fields in experiments by configuring a bosonic Hall system with a molecular BEC in a rotating optical lattice.

References

1. Zhang, L.D., Fu, L.B.: Europhys. Lett. **93**, 30001 (2011)
2. Sachdev, S.: Quantum Phase Transitions. Cambridge University Press, Cambridge (1999)
3. Botet, R., Jullien, R., Pfeuty, P.: Phys. Rev. Lett. 49, 478 (1982); Botet, R., Jullien, R. Phys. Rev. B **28**, 3955 (1983)
4. Dusuel, S., Vidal, J.: Phys. Rev. Lett. 93, 237204 (2004); Dusuel, S., Vidal, J.: Phys. Rev. A 71, 060304(R) (2005); Ribeiro, P., Vidal, J., Mosseri, R.: Phys. Rev. Lett. 99, 050402 (2007); Ribeiro, P., Vidal, J., Mosseri, R.: Phys. Rev. E 78, 021106 (2008); Liberti, G., Piperno, F., Plastina, F. Phys. Rev. A **81**, 013818 (2010)
5. Liu, J., Wu, B., Niu, Q.: Phys. Rev. Lett. **90**, 170404 (2003)
6. Pu, H., Maenner, P., Zhang, W., Ling, H.Y.: Phys. Rev. Lett. **98**, 050406 (2007)
7. Liu, J., Fu, L.B.: Phys. Rev. A **81**, 052112 (2010)
8. Yi, X.X., Huang, X.L., Wang, W.: Phys. Rev. A **77**, 052115 (2008)
9. Oh, S.: Phys. Lett. A 373, 644 (2009); Wu, B., Zhang, Q., Liu, J.: Phys. Lett. A 375, 545 (2011); Sjoqvist, E., Rahaman, R., Basu, U., Basu, B. J. Phys. A: Math. Theor. **43**, 354026 (2010)

10. Carollo, A.C.M., Pachos, J.K.: Phys. Rev. Lett. 95, 157203 (2005); Zhu, S.L.: Phys. Rev. Lett. 96, 077206 (2006); Hamma, A.: Quant-ph/0602091, (2006); Plastina, F., Liberti, G., Carollo, A.: Europhys. Lett. 76, 182 (2006); Chen, G., Li J., Liang, J.Q.: Phys. Rev. A 74, 054101 (2006); Cui, H.T., Li, K., Yi, X.X. Phys. Lett. A **360**, 243 (2006)
11. Fu, L.B., Liu, J.: Ann. Phys. **325**, 2425 (2010)
12. Tikhonenkov, I., et al.: Phys. Rev. A **73**, 043605 (2006)
13. Li, J., et al.: Phys. Rev. A 79, 025602 (2009); Jing, H., Cheng, J., Meystre, P. Phys. Rev. Lett. **99**, 133002 (2007)
14. Liu, J., Fu, L.B.: Phys. Lett. A **370**, 17 (2007)
15. Aharonov, Y., Anandan, J.: Phys. Rev. Lett. 58, 1593 (1987); Wilczek, F., Zee, A.: Phys. Rev. Lett. 52, 1984 (2111); Samuel, J., Bhandari, R.: Phys. Rev. Lett. 60, 2339 (1988); Sjoqvist, E., Pati, A.K., Ekert, A., Anandan, J.S., Ericsson, M., Oi, D.K.L., Vedral, V. Phys. Rev. Lett. **85**, 2845 (2000)
16. Ling, Y.H., et al.: Phys. Rev. A **75**, 033615 (2007)
17. Garrison, J.C., Chiao, R.Y.: Phys. Rev. Lett. 60, 165 (1988); Wu, B., Liu, J., Niu, Q. Phys. Rev. Lett. **94**, 140402 (2005)
18. Berry, M.V.: Fundamental aspects of quantum theory. In: Gorini, V., Frigerio, A. (eds.) NATO ASI Series, vol. 144. Plenum, New York (1986)
19. Simon, B.: Phys. Rev. Lett. **51**, 2167 (1983)
20. Leboeuf, P., Mouchet, A.: J. Phys. A **36**, 2847 (2003)
21. Nunnenkamp, A., Meiser, D., Meystre, P.: New J. Phys. **8**, 88 (2006)
22. Damski, B., Santos, L., Tiemann, E., Lewenstein, M., Kotochigova, S., Julienne, P., Zoller, P.: Phys. Rev. Lett. **90**, 110401 (2003)
23. Li, S.C., Fu, L.B., Liu, J.: Phys. Rev. A **89**, 023628 (2014)
24. Zhou, L., Zhang, W., Ling, H.Y., Jiang, L., Pu, H.: Phys. Rev. A **75**, 043603 (2007)
25. Drobný, G., Jex, I., Bužek, V.: Phys. Rev. A **48**, 569 (1993)
26. Shen, Y.R.: The Principle of Nonlinear Optics. Wiley Interscience, New York (1984)
27. Papp, S.B., Wieman, C.E.: Phys. Rev. Lett. **97**, 180404 (2006)
28. Vardi, A., Yurovsky, V.A., Anglin, J.R.: Phys. Rev. A **64**, 063611 (2001)
29. Pazy, E., Tikhonenkov, I., Band, Y.B., Fleischhauer, M., Vardi, A.: Phys. Rev. Lett. 95, 170403 (2005); Tikhonenkov, I., Pazy, E., I., Band, Y.B., Fleischhauer, M., Vardi, A. Phys. Rev. A **73**, 043605 (2006)
30. Berry, M.V.: Proc. Roy. Soc. A **392**, 45 (1984)
31. Yaffe, L.G.: Rev. Mod. Phys. **54**, 407 (1982)
32. Li, S.C.: J. Phys. B: At. Mol. Opt. Phys. **43**, 205303 (2010)
33. Cui, F., Wu, B.: Phys. Rev. A **84**, 024101 (2011)
34. Wu, B., Liu, J., Niu, Q.: Phys. Rev. Lett. **94**, 140402 (2005)
35. Berry, M.V.: J. Phys. A **18**, 15 (1985)
36. Li, S.C., Liu, J., Fu, L.B.: Phys. Rev. A **83**, 042107 (2011)
37. Santos, G., Foerster, A., Links, J., Mattei, E., Dahmen, S.R.: Phys. Rev. A **81**, 063621 (2010)
38. Li, S.C., Fu, L.B.: Phys. Rev. A **84**, 023605 (2011)
39. Leboeuf, P., Mouchet, A.: J. Phys. A: Math. Gen. **36**, 2847 (2003)
40. Fang, Z., Nagaosa, N., Takahashi, K.S., Asamitsu, A., Mathieu, R., Ogasawara, T., Yamada, H., Kawasaki, M., Tokura, Y., Terakura, K.: Science **302**, 92 (2003)

Chapter 5
Applications of Nonlinear Adiabatic Evolution

Abstract In this chapter, we show selected applications of nonlinear adiabatic evolution in the geometric phase, in tunneling dynamics, and in quantum interference. We introduce the adiabatic geometric phase in a nonlinear coherent coupler and illustrate the nonlinear adiabatic tunneling with four cases, namely, Landau-Zener tunneling, Rosen-Zener tunneling, atom-molecule conversion, and composite adiabatic passage. Adiabatic nonlinear Ramsey interferometry is also discussed.

5.1 Nonlinear Coherent Optical Coupler

In the previous chapters, motivated partly by the studies of Bose-Einstein condensation [1–3], both the adiabatic theorem and the adiabatic geometric phase have been extended to nonlinear systems [4–7]. Here, we further extend these theories to study the effect of nonlinearity on the adiabatic process in coupled waveguide systems, which can also be described by the nonlinear Schrödinger equation. Nonlinear waveguide systems can serve as direct analogies to various other quantum processes [8] and provide ideal models to study nonlinear adiabatic evolution [9].

One of the simplest nonlinear waveguide systems is known as the nonlinear coherent coupler, which consists of two parallel optical waveguides with Kerr nonlinearity [10]. When the two waveguides in the coupler are different, the coupler is called an asymmetric coupler. In this section, we consider the phase mismatch in the asymmetric coupler changing adiabatically with distance, and we calculate the adiabatic geometric phase associated with the supermode of the coupler analytically. We find that the phase is dependent on the input light intensity, and we show the characteristics of the phase at the critical light intensity where different supermodes merge and at the low- and high-intensity limits [9]. Because the nonlinear coherent coupler performs multiple useful functions in optical communications, including power division, power coupling, and switching [11], we expect the geometric phase presented here to have many prospective practical applications.

The model describing the propagation of the laser field inside the coupler can be derived from the standard coupled mode theory [12]. To be clear and self-contained, we first introduce this model briefly. We begin by considering a linearly polarized

J. Liu et al., *Nonlinear Adiabatic Evolution of Quantum Systems*,
https://doi.org/10.1007/978-981-13-2643-1_5

laser field propagating inside the coupler along the $+z$-direction. According to the coupled mode theory, the electric field can be expressed as

$$\boldsymbol{E}(x, y, z, t) = \frac{1}{2}\sum_l A_l(z)\boldsymbol{E}_l(x, y)\mathrm{e}^{i(\beta_l z-\omega t)} + c.c., \tag{5.1}$$

where $c.c.$ denotes the complex conjugate of the term in brackets, ω is the frequency, $\boldsymbol{E}_l$ (with $l = 1, 2$) is the only confined mode of waveguide l, and A_l and $\beta_l > 0$ are the corresponding amplitude and propagation constant, respectively. The mode function $\boldsymbol{E}_l$ satisfies the orthonormalization relation $(\beta_l/(2\omega\mu_0))\int \boldsymbol{E}_l^* \cdot \boldsymbol{E}_{l'} dxdy = \delta_{ll'}$ and the wave equation $\left(\nabla_\perp^2 + (\omega^2/c^2)[1+\chi_l(x, y)]\right)\boldsymbol{E}_l = \beta_l^2\boldsymbol{E}_l$, where $\nabla_\perp^2 = \partial^2/\partial x^2 + \partial^2/\partial y^2$, and the susceptibility distribution is

$$\chi_l(x, y) = \begin{cases} \chi_l, & \text{waveguide } l, \\ 0, & \text{otherwise}. \end{cases} \tag{5.2}$$

The electric field $\boldsymbol{E}$ itself satisfies

$$\left(\nabla^2 + \frac{\omega^2}{c^2}\left[1 + \sum_l \chi_l(x, y)\right]\right)\boldsymbol{E} = \mu_0\frac{\partial^2}{\partial t^2}\boldsymbol{P}_{NL}, \tag{5.3}$$

where $\boldsymbol{P}_{NL}$ is the nonlinear polarization. Because the laser field is linearly polarized, there is only one component of the third-order nonlinear susceptibility, denoted by $\chi_l^{(3)}$, responsible for the Kerr nonlinearity, and the nonlinear polarization

$$\boldsymbol{P}_{NL} = \frac{3}{2}\sum_l \chi_l^{(3)}(x, y)|\boldsymbol{E}|^2\boldsymbol{E} + c.c., \tag{5.4}$$

where the nonlinear susceptibility distribution is

$$\chi_l^{(3)}(x, y) = \begin{cases} \chi_l^{(3)}, & \text{waveguide } l, \\ 0, & \text{otherwise}. \end{cases} \tag{5.5}$$

We now take the scalar product of (5.3) with $\boldsymbol{E}_l^*(x, y)$ and integrate over the entire $x - y$ plane. Moreover, using the slowly varying amplitude approximation $|\beta_l(dA_l/dz)| \gg |d^2A_l/dz^2|$ and assuming that $\int \chi_l^{(3)}(x, y)|\boldsymbol{E}_l|^4 dxdy$ is much larger than the other integrals related to the nonlinearity, we obtain the coupled nonlinear Schrödinger equations [9]

$$i\frac{d}{dz}A_1 + \alpha_1 A_1 + k_{12}\mathrm{e}^{i(\beta_2-\beta_1)z}A_2 + \gamma_1|A_1|^2A_1 = 0, \tag{5.6}$$

$$i\frac{d}{dz}A_2 + \alpha_2 A_2 + k_{21}\mathrm{e}^{-i(\beta_2-\beta_1)z}A_1 + \gamma_2|A_2|^2A_2 = 0, \tag{5.7}$$

where

$$\alpha_1 = \frac{\omega\varepsilon_0}{4}\int \chi_2(x,y)|\boldsymbol{E}_1|^2 dxdy, \tag{5.8}$$

$$\alpha_2 = \frac{\omega\varepsilon_0}{4}\int \chi_1(x,y)|\boldsymbol{E}_2|^2 dxdy, \tag{5.9}$$

$$k_{12} = \frac{\omega\varepsilon_0}{4}\int \chi_1(x,y)\boldsymbol{E}_1^* \cdot \boldsymbol{E}_2 dxdy, \tag{5.10}$$

$$k_{21} = \frac{\omega\varepsilon_0}{4}\int \chi_2(x,y)\boldsymbol{E}_2^* \cdot \boldsymbol{E}_1 dxdy, \tag{5.11}$$

$$\gamma_1 = \frac{3\omega}{4}\int \chi_1^{(3)}(x,y)|\boldsymbol{E}_1|^4 dxdy, \tag{5.12}$$

$$\gamma_2 = \frac{3\omega}{4}\int \chi_2^{(3)}(x,y)|\boldsymbol{E}_2|^4 dxdy. \tag{5.13}$$

Before proceeding further, we comment on (5.6) and (5.7) and the parameters given in (5.8)–(5.13). First, we note that the phase mismatch factors in the coupling terms of (5.6) and (5.7) exist only when $\beta_1 \neq \beta_2$. Second, the parameters α_1 and α_2 serve only to modify β_1 and β_2. Third, from the conservation of the total light intensity $I = |A_1|^2 + |A_2|^2$ in the z-direction, we find $k_{12} = k_{21}^*$. For simplicity, we assume that $k_{12} = k_{21} = k > 0$. Fourth, if $\chi_l^{(3)} > 0$ and, thus, $\gamma_l > 0$, we describe the nonlinear mechanism of waveguide l as being self-focusing; if $\chi_l^{(3)} < 0$ and, thus, $\gamma_l < 0$, we describe the mechanism as being defocusing [13]. In the self-focusing case, the refractive index increases locally with the power, while in the defocusing case, it decreases.

Because the concept of the adiabatic geometric phase originates from quantum mechanics, we need to map the above model into a nonlinear quantum model. Specifically, introducing the nonlinear Hamiltonian

$$H(|A_1|^2, |A_2|^2) = -\begin{pmatrix} \alpha_1 + \gamma_1|A_1|^2 & ke^{i(\beta_2-\beta_1)z} \\ ke^{-i(\beta_2-\beta_1)z} & \alpha_2 + \gamma_2|A_2|^2 \end{pmatrix}, \tag{5.14}$$

we can express the coupled nonlinear Schrödinger equations (5.6) and (5.7) as

$$i\frac{d}{dz}\begin{pmatrix} A_1 \\ A_2 \end{pmatrix} = H(|A_1|^2, |A_2|^2)\begin{pmatrix} A_1 \\ A_2 \end{pmatrix}. \tag{5.15}$$

Here, we note that the evolution of $(A_1, A_2)^T$ in the $+z$-direction corresponds to the time evolution of a nonlinear two-level system in quantum mechanics. The eigenequation of $H(|\bar{A}_1|^2, |\bar{A}_2|^2)$ reads

$$\mu\begin{pmatrix} \bar{A}_1 \\ \bar{A}_2 \end{pmatrix} = H(|\bar{A}_1|^2, |\bar{A}_2|^2)\begin{pmatrix} \bar{A}_1 \\ \bar{A}_2 \end{pmatrix}, \tag{5.16}$$

where μ is the eigenvalue and $(\bar{A}_1, \bar{A}_2)^T$ is the eigenstate or the supermode of the coupler [11].

For convenience and without loss of generality, we further write

$$\begin{pmatrix} A_1 \\ A_2 \end{pmatrix} = \sqrt{I} \begin{pmatrix} \cos(\theta/2)\mathrm{e}^{-i\phi_1} \\ \sin(\theta/2)\mathrm{e}^{-i\phi_2} \end{pmatrix} = \sqrt{I} \begin{pmatrix} \cos(\theta/2) \\ \sin(\theta/2)\mathrm{e}^{-i\phi} \end{pmatrix} \mathrm{e}^{-i\phi_1}, \tag{5.17}$$

where $0 \le \theta \le \pi$, $\phi = \phi_2 - \phi_1$, and ϕ_1 has been decoupled as the overall phase. From (5.17), we know that the state of the system, except for an overall phase, can be denoted by (θ, ϕ). Because θ and ϕ span a unit sphere, called the Poincaré sphere [11, 12, 14], the evolution of the system without the overall phase corresponds to the movement of the system on the Poincaré sphere. Introducing $\beta = \beta_2 - \beta_1$ and $\alpha = \alpha_2 - \alpha_1$ and substituting (5.17) into (5.15), we obtain

$$\frac{d\theta}{dz} = -2k \sin(\phi - \beta z), \tag{5.18}$$

$$\frac{d\phi}{dz} = -\alpha - 2k \cot\theta \cos(\phi - \beta z) - I\gamma_2 + I(\gamma_1 + \gamma_2)\cos^2(\theta/2), \tag{5.19}$$

$$\frac{d\phi_1}{dz} = -\alpha_1 - k \tan(\theta/2)\cos(\phi - \beta z) - I\gamma_1 \cos^2(\theta/2). \tag{5.20}$$

Similar to (5.17), we write $(\bar{A}_1, \bar{A}_2)^T = (\cos(\bar{\theta}/2), \sin(\bar{\theta}/2)\mathrm{e}^{-i\bar{\phi}})^T$ so that we can denote the supermode by $(\bar{\theta}, \bar{\phi})$. From (5.16), we have

$$\bar{\phi} = \beta z \text{ and } \bar{\phi} = \beta z + \pi, \tag{5.21}$$

$$\pm 2k \cot\bar{\theta} = -\alpha - I\gamma_2 + I(\gamma_1 + \gamma_2)\cos^2(\bar{\theta}/2), \tag{5.22}$$

$$\mu = -\alpha_1 \mp k \tan(\bar{\theta}/2) - I\gamma_1 \cos^2(\bar{\theta}/2). \tag{5.23}$$

Here and below, the upper sign corresponds to $\bar{\phi} = \beta z$, and the lower sign corresponds to $\bar{\phi} = \beta z + \pi$.

Introducing $t = \tan(\bar{\theta}/2)$, we can rewrite (5.22) as

$$t^4 \mp \frac{\alpha + I\gamma_2}{k} t^3 \mp \frac{\alpha - I\gamma_1}{k} t - 1 = 0. \tag{5.24}$$

Because $0 \le \bar{\theta} \le \pi$, only $t \ge 0$ corresponds to the supermode of the coupler. Note that if t_0 is a solution to (5.24) for the upper sign, then $-t_0$ is a solution for the lower sign. Therefore, we can find all supermodes by solving (5.24) for either sign and changing the signs of the minus solutions. Actually, without having to solve (5.24), we can analytically prove that only when

$$\left[2\alpha + I(\gamma_2 - \gamma_1)\right]^{2/3} + (4k)^{2/3} = \left[I(\gamma_1 + \gamma_2)\right]^{2/3} \tag{5.25}$$

does Eq. (5.24) have three solutions for either sign. If the left-hand side of (5.25) is greater (less) than the right-hand side, then (5.24) has two (four) solutions for either sign. This result indicates that (5.25) determines the parameter region where the number of supermodes changes.

We consider an asymmetric coupler so that the phase mismatch βz is nonvanishing and the supermode changes with z according to (5.21). Mathematically, the supermode corresponds to the fixed point on the Poincaré sphere; therefore, the change of the supermode corresponds to the movement of the fixed point on the sphere. Moreover, we assume that β is small enough that—according to the nonlinear version of the adiabatic evolution condition [4]—the system, which is in a fixed point at $z = 0$, can remain in this fixed point and thus also moves on the Poincaré sphere as z increases. When z reaches $2\pi/\beta$, the system returns to its initial position on the Poincaré sphere, and the overall phase ϕ_1 acquires an increasing component that consists of an adiabatic geometric phase Φ in addition to the dynamical phase. To obtain the expression for Φ, we need to use the method introduced in [6] to separate the Φ-related term from (5.20). This process is equivalent to calculating Φ as the difference between the overall phase and the dynamical phase [15]. To proceed, we first note that because β is small but finite, the system fluctuates around the supermode during the evolution process, i.e., $\phi = \bar{\phi} + \delta\phi$ and $\theta = \bar{\theta} + \delta\theta$, where $\delta\phi \sim \delta\theta \sim O(\beta)$. Then, using (5.18)–(5.23) and ignoring the terms $\sim O(\beta^2)$, we have

$$\frac{d\bar{\theta}}{dz} = \mp 2k\delta\phi, \tag{5.26}$$

$$\frac{d\bar{\phi}}{dz} = \frac{1}{2}\left[\frac{\pm 4k}{\sin^2\bar{\theta}} - I(\gamma_1 + \gamma_2)\sin\bar{\theta}\right]\delta\theta, \tag{5.27}$$

$$\frac{d\phi_1}{dz} = \mu - \frac{1}{2}\left[\frac{\pm k}{\cos^2(\bar{\theta}/2)} - I\gamma_1\sin\bar{\theta}\right]\delta\theta. \tag{5.28}$$

Combining (5.27) and (5.28), we have

$$\frac{d\phi_1}{dz} = \mu - \frac{\beta}{2}\left[1 - \frac{\cos\bar{\theta} \mp (I/4k)(\gamma_2 - \gamma_1)\sin^3\bar{\theta}}{1 \mp (I/4k)(\gamma_1 + \gamma_2)\sin^3\bar{\theta}}\right]. \tag{5.29}$$

Integrating the zero-order term μ over z from 0 to $2\pi/\beta$, we obtain the dynamical phase. Similarly, integrating the Φ-related first-order term, we obtain the adiabatic geometric phase [9]

$$\Phi = -\pi\left[1 - \frac{\cos\bar{\theta} \mp (I/4k)(\gamma_2 - \gamma_1)\sin^3\bar{\theta}}{1 \mp (I/4k)(\gamma_1 + \gamma_2)\sin^3\bar{\theta}}\right]. \tag{5.30}$$

From (5.30), we know that the geometric phase Φ depends on the total light intensity I. For the low-intensity limit $I \to 0$, the nonlinear effect is negligible. In this case, (5.30) reduces to $\Phi = -\pi(1 - \cos\bar{\theta})$, which denotes half of the area on the Poincaré sphere enclosed by the evolution loop of the system except for a sign difference due

Table 5.1 Supermodes and Φ values in the high-intensity limit $I \to \infty$ with $\gamma_1\gamma_2 < 0$

$\bar{\phi}(\gamma_1 > 0, \gamma_2 < 0)$	$\bar{\phi}(\gamma_1 < 0, \gamma_2 > 0)$	$\cos\bar{\theta} \to$	$\Phi \to$
βz	$\beta z + \pi$	1	0
$\beta z + \pi$	βz	-1	-2π

Table 5.2 Supermodes and Φ values in the high-intensity limit $I \to \infty$ with $\gamma_1\gamma_2 > 0$

$\bar{\phi}(\gamma_1, \gamma_2 > 0)$	$\bar{\phi}(\gamma_1, \gamma_2 < 0)$	$\cos\bar{\theta} \to$	$\Phi \to$
βz	βz	$\frac{\gamma_2-\gamma_1}{\gamma_1+\gamma_2}$	$-\frac{2\gamma_1}{\gamma_1+\gamma_2}\pi$
$\beta z + \pi$	$\beta z + \pi$	$\frac{\gamma_2-\gamma_1}{\gamma_1+\gamma_2}$	$-\frac{2\gamma_1}{\gamma_1+\gamma_2}\pi$
βz	$\beta z + \pi$	1	0
βz	$\beta z + \pi$	-1	-2π

Table 5.3 Supermodes and Φ values when $\alpha = 0$ and $\gamma_1 = \gamma_2 > 0$

$\bar{\phi}$	$\cos\bar{\theta}$	Φ
$\beta z + \pi$	0	$-\pi$
βz	0	$-\pi$ (unless $I\gamma_1 = 2k$)
βz	$[1 - (\frac{2k}{I\gamma_1})^2]^{\frac{1}{2}}$	$-\pi + \pi[1 - (\frac{2k}{I\gamma_1})^2]^{-\frac{1}{2}}$
βz	$-[1 - (\frac{2k}{I\gamma_1})^2]^{\frac{1}{2}}$	$-\pi - \pi[1 - (\frac{2k}{I\gamma_1})^2]^{-\frac{1}{2}}$

to the definition of the overall phase. Noting that this result is in accordance with the result obtained in linear systems [16–18], we conclude that the geometric phase Φ is a nonlinear extension of its linear counterpart. In the low-intensity limit, from (5.22), we have $\cos\bar{\theta} \to \mp\alpha/\sqrt{\alpha^2 + 4k^2}$, and thus, $\Phi \to -\pi(1 \pm \alpha/\sqrt{\alpha^2 + 4k^2})$. In the high-intensity limit $I \to \infty$, the nonlinear effect is dominant. Then, using (5.22) to obtain $\cos\bar{\theta}$, we find that (5.30) still reduces to $\Phi = -\pi(1 - \cos\bar{\theta})$. That is, the geometric meaning of Φ remains the same as in the low-intensity limit.

To illustrate the geometric phase Φ and the supermode in the high-power-intensity limit, we list these components in Tables 5.1 and 5.2. We find that the supermodes in Table 5.1 and in the last two lines of Table 5.2 confine the laser field in one of the two waveguides. In this case, $\Phi = 0$ or -2π trivially. The supermodes in the first two lines of Table 5.2 have an intensity distribution determined by the ratio between γ_1 and γ_2. The values of Φ are also determined by this ratio.

In contrast to the geometric phases in linear systems, a remarkable feature of Φ lies in its divergence when

$$\sin^3\bar{\theta} = \pm\frac{4k}{I(\gamma_1 + \gamma_2)}. \tag{5.31}$$

From (5.22) and (5.31), we obtain

Table 5.4 Supermodes and Φ values when $\alpha = 0$ with $\gamma_1 = \gamma_2 < 0$

$\bar{\phi}$	$\cos\bar{\theta}$	Φ
βz	0	$-\pi$
$\beta z + \pi$	0	$-\pi$ (unless $I\gamma_1 = -2k$)
$\beta z + \pi$	$[1 - (\frac{2k}{I\gamma_1})^2]^{\frac{1}{2}}$	$-\pi + \pi[1 - (\frac{2k}{I\gamma_1})^2]^{-\frac{1}{2}}$
$\beta z + \pi$	$-[1 - (\frac{2k}{I\gamma_1})^2]^{\frac{1}{2}}$	$-\pi - \pi[1 - (\frac{2k}{I\gamma_1})^2]^{-\frac{1}{2}}$

$$\cos^3\bar{\theta} = \frac{2\alpha + I(\gamma_2 - \gamma_1)}{I(\gamma_1 + \gamma_2)}. \tag{5.32}$$

Combining (5.31) and (5.32) yields (5.25) exactly. Consequently, only when (5.25) holds can the geometric phase Φ diverge. Noting that (5.25) determines the parameter region where the number of supermodes changes, we find that the divergence of Φ is caused by the merging of the supermodes. Actually, in addition to the divergence condition, from (5.31) and (5.32), we can determine the merged supermode completely, which indicates that the geometric phase Φ characterizes the supermode precisely.

To illustrate the divergence of Φ when the supermodes merge, we take $\alpha = 0$ and $\gamma_1 = \gamma_2$, and we list all supermodes and Φ values in Tables 5.3 and 5.4. Note that the supermodes and Φ values in the last two lines of these two tables exist only when $|I\gamma_1| > 2k$ and that the geometric phase Φ values in the second lines diverge when $|I\gamma_1| = 2k$. At the critical intensity $I = 2k/|\gamma_1|$, the supermodes in the last three lines of these two tables have the same $\bar{\phi}$ and $\cos\bar{\theta}$ and thus merge together. On the other hand, at the same critical intensity, the geometric phase Φ values in the last three lines diverge.

From the perspective of spontaneous symmetry breaking [19], the supermodes in the first two lines of Tables 5.3 and 5.4 are symmetric supermodes, and the supermodes in the last two lines are broken supermodes. As the total intensity I increases adiabatically, the system, which is initially in a symmetric supermode, can ultimately settle in a broken supermode. This process is actually a continuous phase transition that occurs when $|I\gamma_1| = 2k$. The divergence of Φ can be interpreted as the signal of the phase transition. Recently, a relation between geometric phases and phase transitions was proposed [20]. The geometric phase Φ in the continuous phase transition provides a paradigm for this relation, and the general correspondence between the divergence of Φ and the merging of the supermode can be regarded as an extension of this relation.

To illustrate the geometric phase Φ in other cases, we need to calculate the supermode and Φ numerically. As an example, in Fig. 5.1, we show the changes of $\cos\bar{\theta}$ and Φ as γ_2 tends to $\gamma_1 = 1$ with $\alpha = 0$ and $k = 1$. From this figure, we can confirm the following features of $\cos\bar{\theta}$ and Φ. First, in the low-intensity limit $I \to 0$, we have $\cos\bar{\theta} \to 0$ and $\Phi \to -\pi$ for both $\bar{\phi} = \beta z$ and $\bar{\phi} = \beta z + \pi$, as given before. Second, when the intensity I is large enough, $\cos\bar{\theta}$ and Φ tend to the values given in

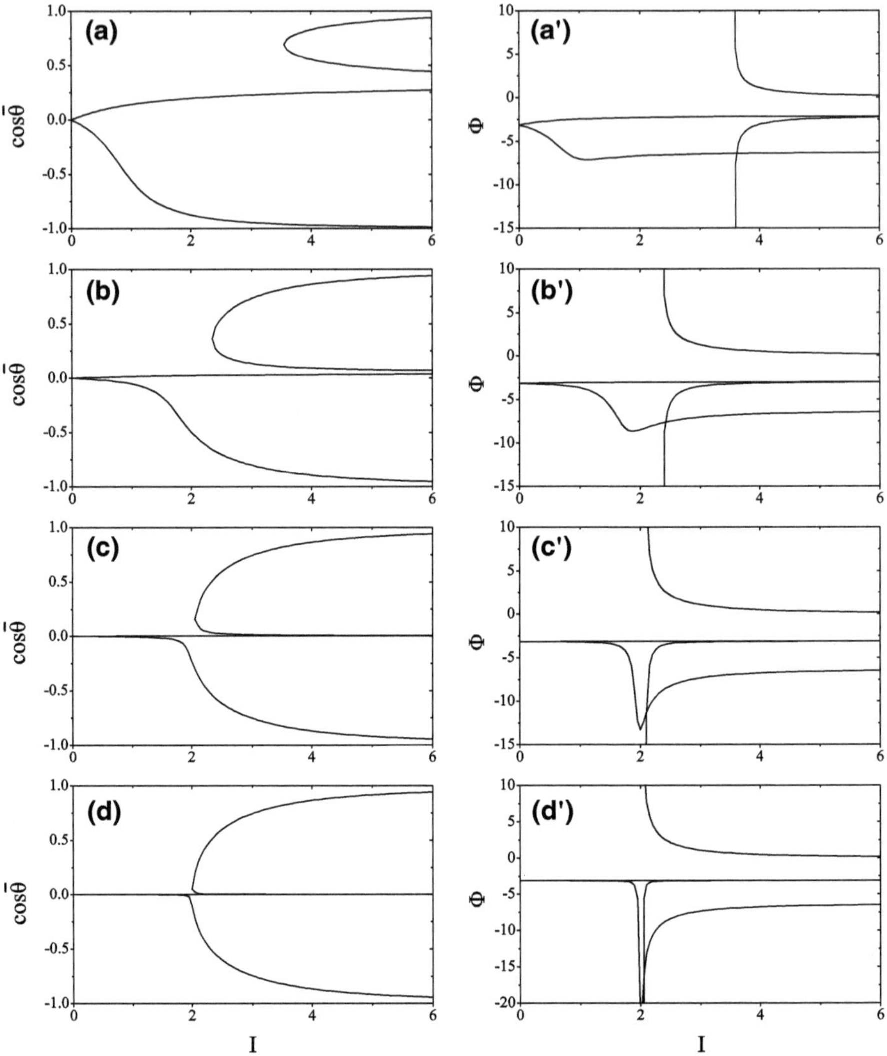

Fig. 5.1 Changes of $\cos\bar{\theta}$ (left panels) and Φ (right panels) as γ_2 tends to $\gamma_1 = 1$ with $\alpha = 0$ and $k = 1$. **a, a'** $\gamma_2 = 2$, **b, b'** $\gamma_2 = 1.1$, **c, c'** $\gamma_2 = 1.01$, **d, d'** $\gamma_2 = 1.001$

Table 5.2. Third, when two supermodes merge at the critical intensity obtained from (5.25), the geometric phase Φ diverges. Fourth, when γ_2 tends to γ_1, $\cos\bar{\theta}$ and Φ tend to the values given in Table 5.3.

5.2 Nonlinear Landau-Zener Tunneling

5.2.1 Two-Level System

It is common in the study of quantum systems to consider only a finite number of energy levels that are strongly coupled. The special case of two coupled levels is of enormous practical interest, and a vast amount of literature has been devoted to the dynamical properties of such two-level systems [21]. One of the interesting phenomena is the Landau-Zener tunneling between energy levels. As a basic physical process [22], this effect has found wide applications in various systems, such as current-driven Josephson junctions [23], atoms in accelerating optical lattices [24], and field-driven superlattices [25].

A nonlinear two-level system, in which the level energies depend on the occupation of the levels, may arise in a mean-field treatment of a many-body system where the particles predominantly occupy two energy levels. For example, such a model arises in the study of the motion of a small polaron [26], a Bose-Einstein condensate (BEC) in a double-well potential [27–29] or in an optical lattice [30, 31], or a small-capacitance Josephson junction where the charging energy may be important. In contrast to the linear case, the dynamical property of a nonlinear two-level model is far from being fully understood, and many novel features have been revealed [2, 32], including the discovery of a nonzero Landau-Zener tunneling probability even in the adiabatic limit when the nonlinear parameter C exceeds a critical value V.

In this section, we present an analytic discussion on nonlinear Landau-Zener tunneling. For the behavior near the critical point $C = V$, we find that the adiabatic tunneling probability between the two energy levels rises as a $3/2$ power law of the function $C/V - 1$. Below the critical point, the tunneling probability as a function of the sweep rate α follows an exponential law (as in the linear case) but with the exponent modified due to the nonlinearity. The explicit expression of the modification factor is obtained analytically, is found to decrease monotonically with the nonlinear parameter and tends to zero at the critical point, indicating the breakdown of the exponential law. Indeed, our analysis shows that the exponential law breaks down at the critical point and transforms into a $3/4$ power law. Beyond the critical regime, i.e., for $C > V$, we employ the stationary phase method and obtain a closed-form solution of the nonlinear tunneling probability. This solution is compared with the numerical solution by integrating the Schrödinger equation; the two solutions exhibit satisfactory agreement for a wide range of parameters [3].

Our model consists of two levels (as in the standard Landau-Zener model) but with an additional energy difference depending on the population in the levels. This configuration is described by the following Hamiltonian [2]:

$$H(\gamma) = \begin{pmatrix} \frac{\gamma}{2} + \frac{C}{2}(|b|^2 - |a|^2) & \frac{V}{2} \\ \frac{V}{2} & -\frac{\gamma}{2} - \frac{C}{2}(|b|^2 - |a|^2) \end{pmatrix}, \tag{5.33}$$

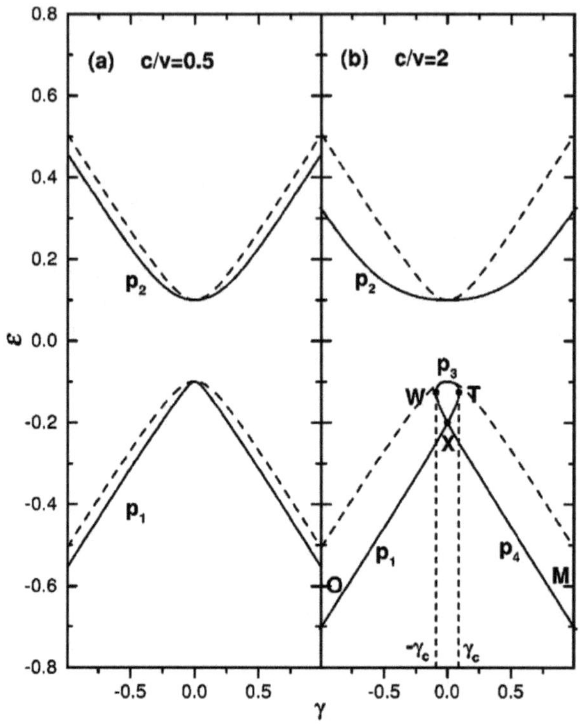

Fig. 5.2 Adiabatic energy levels (solid lines) for two typical nonlinear cases: **a** $C = 0.1$, $V = 0.2$; **b** $C = 0.4$, $V = 0.2$. The dashed lines are for the linear case ($C = 0$). The corresponding eigenstates are the fixed points P_i ($i = 1, \cdots, 4$) of the H_e system (5.36) as shown in (**b**): $OXT \to P_1$, $MXW \to P_4$, $WT \to P_3$. Only P_3 is an unstable saddle point; the others are stable elliptic points

where a and b are the probability amplitudes. The Hamiltonian is characterized by three parameters: the coupling V between the two levels, the level bias γ, and the nonlinear parameter C describing the level energy dependence on the populations. The amplitudes a and b satisfy the Schrödinger equation,

$$i\frac{d}{dt}\begin{pmatrix} a \\ b \end{pmatrix} = H(\gamma)\begin{pmatrix} a \\ b \end{pmatrix}, \tag{5.34}$$

which conserves the total probability $|a|^2 + |b|^2$, which is set to 1.

We wish to examine the nonlinear Landau-Zener tunneling, i.e., how the system evolves when the level bias γ changes with time as $\gamma = \alpha t$. We define a sweep rate α. In this section, we focus on the adiabatic limit, that is, the case in which the sweep rate α tends to zero.

As in the linear model, it is useful to find the adiabatic levels $\varepsilon(\gamma)$ by diagonalizing the Hamiltonian (5.33). It is readily found that in the nonlinear model, there are two eigenvalues when $C < V$, while there can be four eigenvalues when $C > V$, as demonstrated in Fig. 5.2. At $C/V = 2$ (Fig. 5.2b), because of the four eigenvalues, a loop appears at the tip of the lower level in the regime $-\gamma_c \leq \gamma \leq \gamma_c$, where

$$\gamma_c = (C^{2/3} - V^{2/3})^{3/2}. \tag{5.35}$$

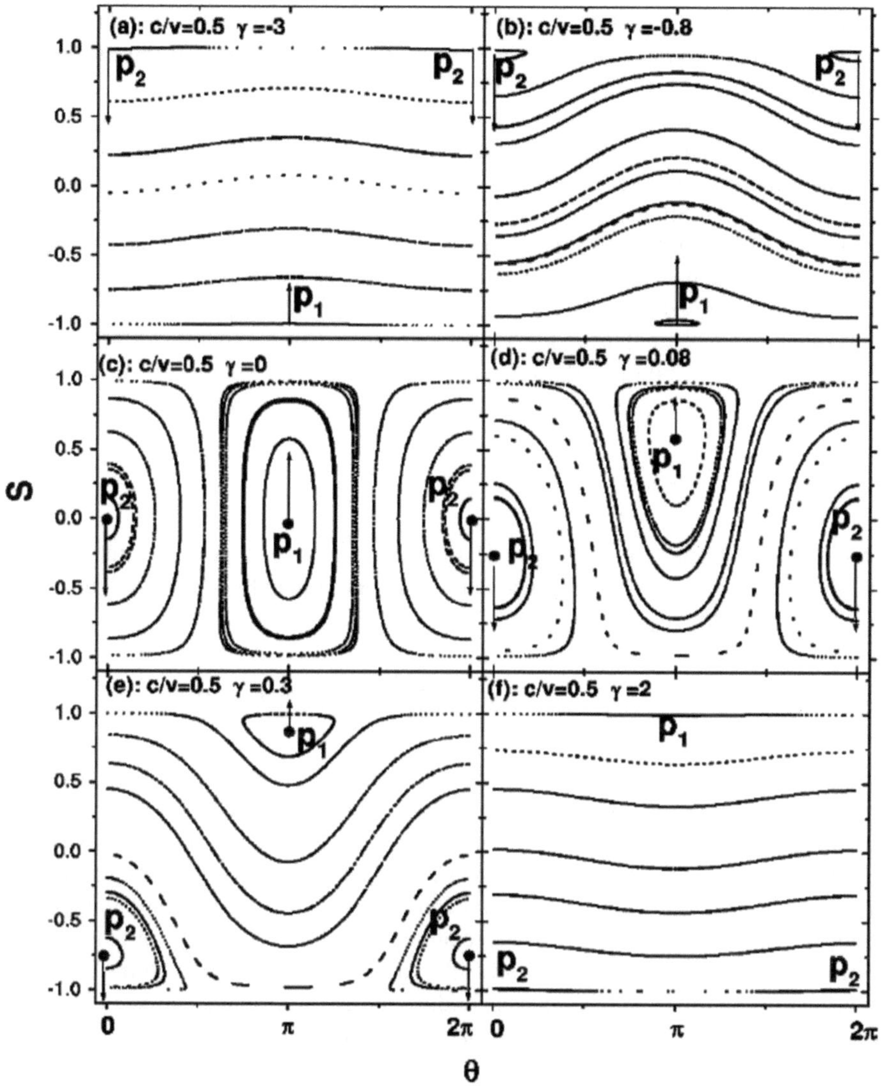

Fig. 5.3 Evolution of the phase-space motions of the Hamiltonian system H_e at $C/V = 0.5$ as γ changes adiabatically. The arrows indicate the shifting direction of the fixed points P_i as γ increases. The closed curves are the periodic trajectories. In this case, no collision between fixed points occurs, implying zero adiabatic tunneling probability

The corresponding eigenstates are not orthogonal to each other for finite γ but become so in the limits of $\gamma \to \pm\infty$, where $\varepsilon \to \pm|\gamma|/2$. For instance, at the lower level, we have $(a, b) \to (1, 0)$ at $\gamma \to -\infty$ and $(a, b) \to (0, 1)$ at $\gamma \to +\infty$.

The direct consequence of the loop structure in Fig. 5.2b, as first discussed in [2], is that as a quantum state moves along the lower lever to the singular point T, no options remain for further movement except to jump to the upper and lower levels. As a result, the nonlinear Landau-Zener tunneling is not zero even in the adiabatic limit $\alpha \to 0$. The underlying mechanism of this interesting phenomenon is revealed with an equivalent classical Hamiltonian, where the nonzero adiabatic tunneling probability is viewed as the result of collision between fixed points.

With $a = |a|e^{i\theta_a}$ and $b = |b|e^{i\theta_b}$, we introduce the population difference $s = |b|^2 - |a|^2$ and the relative phase $\theta = \theta_b - \theta_a$. In terms of s and θ, the nonlinear two-level system is expressed as a classical Hamiltonian system [28, 32],

$$H_e(s, \theta, \gamma) = \frac{C}{2}s^2 + \gamma s - V\sqrt{1 - s^2}\cos\theta, \tag{5.36}$$

which has the form of a Josephson Hamiltonian. The fixed points of the classical Hamiltonian correspond to the eigenstates of the nonlinear two-level system and are given by the following equations:

$$\theta^* = 0, \pi, \quad \gamma + Cs^* + \frac{Vs^*}{\sqrt{1 - (s^*)^2}}\cos\theta^* = 0. \tag{5.37}$$

The number of fixed points depends on the nonlinear parameter C. For weak nonlinearity, $C/V < 1$, there exist only two fixed points (P_1 and P_2 in Fig. 5.3, corresponding to the maximum and minimum of the classical Hamiltonian). These points are elliptic points, each being surrounded by closed (elliptic) orbits. The fixed points are located on the lines of $\theta^* = \pi$ and 0, implying that the two corresponding eigenstates of the two-level system have a relative phase of π. As the level bias changes from $\gamma = -\infty$ to $+\infty$, P_1 moves smoothly along the line $\theta^* = \pi$ from the bottom ($s = -1$) to the top ($s = +1$), corresponding to the lower energy level in Fig. 5.2a; the other point, P_2, moves from the top to the bottom, corresponding to the upper level.

For stronger nonlinearity, $C/V > 1$, two more fixed points appear in the window $-\gamma_c < \gamma < \gamma_c$. As shown in Fig. 5.4c–e, both of the new fixed points lie on the line $\theta^* = \pi$, one being elliptic (P_4) and the other being hyperbolic (P_3) as a saddle point of the classical Hamiltonian. One of the original fixed points, P_2, still moves smoothly with γ, corresponding to the upper adiabatic level in Fig. 5.2b. The other point, P_1, moves smoothly up to $\gamma = \gamma_c$, where it collides with P_3, corresponding to the branch OXT of the lower level in Fig. 5.2b. The new elliptic point P_4, created at $\gamma = -\gamma_c$ together with P_3, moves up to the top, corresponding to the branch WXM of the lower level. The hyperbolic point P_3 moves down away from its partner P_4 after creation and is annihilated with P_1 at $\gamma = \gamma_c$, corresponding to the top branch WT of the lower level. The collision between P_1 and P_3 leads to nonzero adiabatic

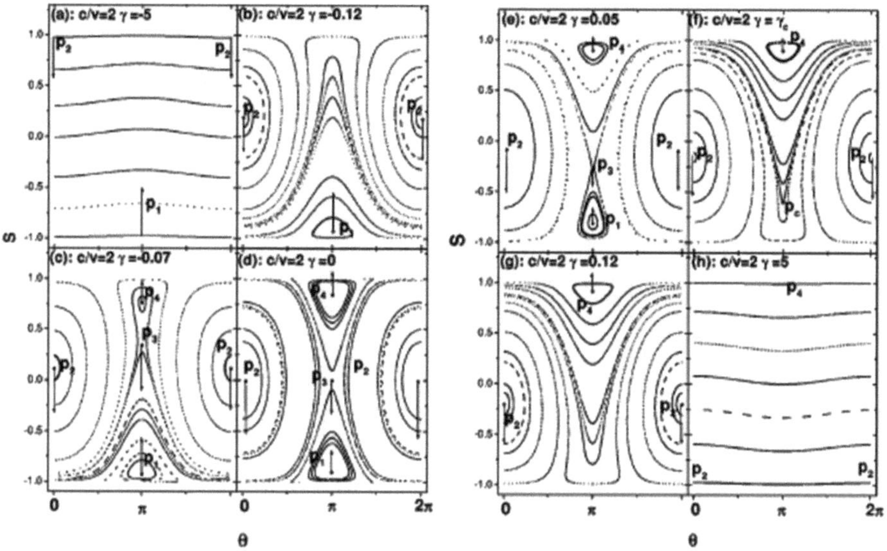

Fig. 5.4 Evolution of the phase-space motions of the Hamilton H_e system at $C/V = 2$ as γ changes adiabatically. The arrows refer to the moving directions of the fixed points as γ increases. In this case, the fixed points P_2 and P_3 collide at the singular point γ_c and form a homoclinic orbit with nonzero action. This jump of the action leads to nonzero adiabatic tunneling probability

tunneling from the lower level to the upper level, which is determined by the ultimate fate of the fixed point P_1.

For adiabatic changes of the level bias γ, a closed orbit in the classical dynamics remains closed, and the action

$$I = \frac{1}{2\pi} \oint s d\theta \tag{5.38}$$

stays invariant in time according to the classical adiabatic theorem [33]. The change of γ is adiabatic as long as the relative change of γ in a period of the orbit is small. The action equals the phase-space area enclosed by the closed orbit and is therefore zero for a fixed point. Since the closed orbits surrounding an elliptic fixed point all have finite periods T, these orbits evolve with the area of each fixed in time. We thus expect an elliptic fixed point to remain a fixed point during the adiabatic change of the level bias γ. For the case of $C/V < 1$, the two fixed points (both elliptic) evolve adiabatically throughout the entire sweeping of γ, implying the absence of a transition between the eigenstates in the adiabatic limit. This condition is still true for the fixed point P_2 in the case of $C/V < 1$, meaning that a state starting from the upper level remains in the upper level.

The adiabaticity is broken, however, when P_1 collides with the hyperbolic fixed point P_3 to form a homoclinic orbit where the period T diverges. Nevertheless, the classical "particle" remains on this orbit because the orbit is surrounded from both outside and inside by closed orbits of finite periods, which form barriers to prevent the particle from escaping. After this collision, the homoclinic orbit turns into an ordinary closed orbit of finite period and evolves adiabatically for $\gamma > \gamma_c$ according to the rule of constant action, which is now nonzero. This orbit ultimately evolves into a straight line of constant s.

With these observations, we can obtain the tunneling probability in the adiabatic limit [3]:

$$\Gamma_{ad} = \frac{1}{2} I(s_c) = \frac{1}{4\pi} \oint s(\theta, E_c) d\theta, \tag{5.39}$$

where

$$s_c = -\sqrt{1 - (V/C)^{2/3}}, \tag{5.40}$$

and

$$E_c = \frac{C}{2} s_c^2 + \gamma_c s_c - V\sqrt{1 - s_c^2}. \tag{5.41}$$

The above analysis is consistent with the nonlinear hysteresis phenomenon presented in [32], where a similar formula for adiabatic tunneling probability was obtained.

The adiabatic tunneling probability can be evaluated analytically in the critical region of $\delta = C/V - 1 \to 0$. The singular point of the level bias is found to leading order as follows:

$$\gamma_c \simeq V(\frac{2}{3}\delta)^{3/2}. \tag{5.42}$$

The homoclinic orbit is confined near the critical point, with its top at

$$s_t \simeq s_c + \sqrt{6\delta}. \tag{5.43}$$

We expand the classical Hamiltonian to leading orders of $(s - s_c)$ and $(\theta - \pi)$ and find

$$\theta - \pi \simeq \sqrt{\frac{2\gamma_c(s - s_c)}{V}} + \frac{1}{2}\sqrt{\frac{2\gamma_c}{V}}(s - s_c)^{3/2}. \tag{5.44}$$

From the area of this orbit, the adiabatic tunneling probability for this limiting case is found to be

$$\Gamma_{ad} = \frac{1}{2\pi} \int_{s_c}^{s_t} (\theta - \pi) ds = \frac{4}{3\pi} \delta^{3/2}. \tag{5.45}$$

Clearly, both Γ_{ad} and its first-order derivative are continuous at the critical point. However, its second-order derivative is discontinuous.

In the linear case of $C = 0$, an exact formula exists that prescribes an exponential dependence of the tunneling probability on the sweep rate [22]:

$$\Gamma_{lz} = \exp\left(-\frac{\pi V^2}{2\alpha}\right). \tag{5.46}$$

It is interesting to consider how this exponential law is changed due to the nonlinearity. We focus first on the near-adiabatic case (i.e., $\alpha \neq 0$ and $\alpha \ll 1$).

For this purpose, we need to investigate the evolution of the fixed point P_1 in addition to the nearby periodic orbits by introducing the angle variable ϕ, which is the canonical conjugate of the action variable I. As in the adiabatic case considered above, the transition probability is still given by the increment of the action, i.e., $\Gamma = \frac{1}{2}\Delta I$. According to the standard theory on the nonadiabatic correction [33], we have

$$\Delta I = \int_{-\infty}^{+\infty} R(I, \phi) \frac{d\gamma}{dt} \frac{d\phi}{\dot{\phi}}, \tag{5.47}$$

where $R(I, \phi)$ is the periodic function of ϕ with zero average and is related to the generating function of the canonical transformation from variables (s, θ) to (I, ϕ). The concrete form of the function R is unimportant in the following discussion.

To evaluate the above integral, we need to express $\dot{\phi}$ as a function of ϕ itself. In the near-adiabatic limit, the change of the angle variable is equal to the frequency of the fixed point P_1, i.e., $\dot{\phi} = \omega^*$. The frequency can be calculated by linearizing the equations of motion (5.36) near the fixed point (5.37),

$$\omega^* = V\left(\frac{1}{1-(s^*)^2} - \frac{C}{V}\sqrt{1-(s^*)^2}\right)^{1/2}. \tag{5.48}$$

On the other hand, by substituting $\theta^* = \pi$ into Eq. (5.37) and differentiating this equation with respect to time, we have

$$\frac{dt}{ds^*} = \frac{V}{\alpha}\left(\frac{1}{[1-(s^*)^2]^{3/2}} - \frac{C}{V}\right). \tag{5.49}$$

Combining these equations, we can relate s^* to ϕ and thus express $\dot{\phi}$ as a function of ϕ itself.

The principal contribution to the integral comes from the neighborhood of the singularities of the integrand, which are the zeros of the frequency $\dot{\phi} = \omega^*(\gamma)$. These zero points are easily found from Eq. (5.48) as follows:

$$s_0^* = [1 - (V/C)^{2/3}]^{1/2}. \tag{5.50}$$

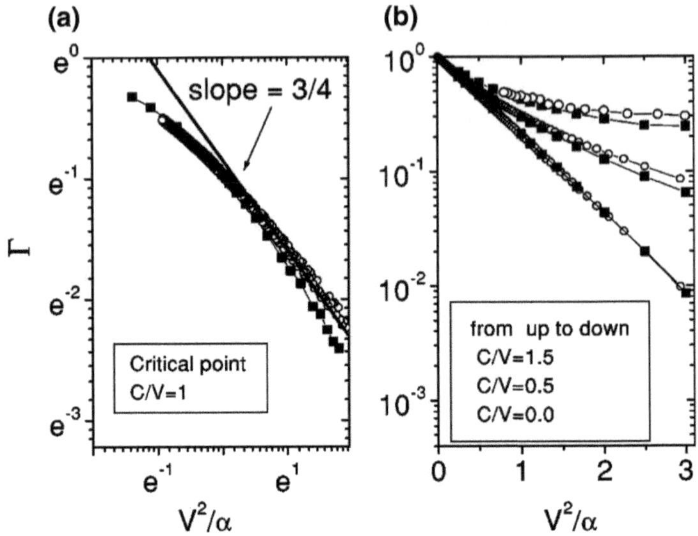

Fig. 5.5 Dependence of the tunneling probability on the scaled sweep rate V^2/α **a** for $C/V = 1$ and **b** other values of C/V. In (**b**), we observe a clear breakdown of the exponential law for $C/V > 1$. The open circles are obtained with the integration of Eq. (5.34); the solid squares are the numerical results of a Bose-Einstein condensate in an accelerating optical lattice, where α is the acceleration

The integral (5.47) is exponentially small if there are no real singularities and becomes a power law in the sweep rate if there is a singularity on the real axis.

We consider the case of critical nonlinearly, $C/V = 1$, for which the singular point occurs at $s^* = 0$. Near this point, we find from Eq. (5.49) that $\omega^* \simeq \sqrt{3/2}Vs^*$ and $\phi \simeq (1/4)(3/2)^{3/2}(V^2/\alpha)(s^*)^4$. Then, we have an approximate relation $\omega^* \sim \alpha^{1/4}\phi^{1/4}$ near the singularity. Substituting these expressions back into Eq. (5.47) and using the fact that $\partial R/\partial\phi$ is independent of α, we find a power-law behavior for the tunneling probability,

$$\Gamma \sim \alpha^{3/4}. \tag{5.51}$$

This power law, which indicates a sharp change of tunneling behavior beyond the critical regime $C = V$, is verified by our numerical calculations (Fig. 5.5a).

We shift our attention to nonadiabatic tunneling for subcritical nonlinearity, $C < V$, where the zeros of the frequency ω^* are complex. The principal contribution to the integral (5.47) comes from the neighborhood of this point, and the integral can be evaluated by deforming the contour of integration into the complex plane [33]. The tunneling probability is found to be exponential,

$$\Gamma \sim \exp\left(-q\frac{\pi V^2}{2\alpha}\right), \tag{5.52}$$

where the factor in the exponent is given by

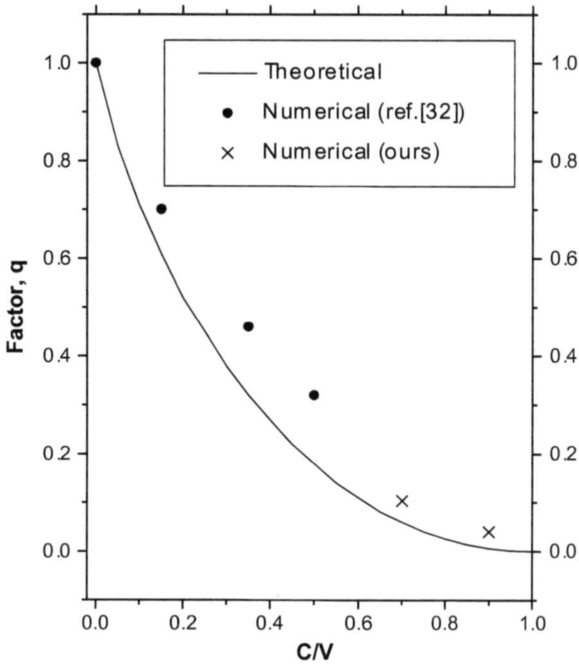

Fig. 5.6 Dependence of the factor q on C/V

$$q = \frac{4}{\pi} \int_0^{\sqrt{(V/C)^{2/3}-1}} (1+x^2)^{1/4} \left(\frac{1}{(1+x^2)^{3/2}} - \frac{C}{V} \right)^{3/2} dx. \tag{5.53}$$

For the linear case $C = 0$, the factor q is exactly unity, consistent with the standard Landau-Zener formula (5.46). For the nonlinear case, $C/V > 0$, this factor becomes less than one, showing the enhancement effect on the nonadiabatic tunneling. As C/V increases to 1, the critical point, this factor vanishes, signaling the breakdown of the exponential law. Near the critical point $C/V = 1$, we have the approximate expression $q \simeq \frac{3}{4}\sqrt{2/3}(1 - C/V)^2$, i.e., the factor converges to zero with a square power law.

With numerical integration of the nonlinear Schrödinger equation (5.34), we show in Fig. 5.5 the sweep rate α dependence of the tunneling probability, where the slope of the curve tends to be zero for $C/V > 1$, clearly indicating the breakdown of the exponential law. We read the factor q from the slope and compare it with our analytical results in Fig. 5.6, finding reasonably good agreement.

We now discuss the nonlinear Landau-Zener tunneling beyond the critical regime and derive the tunneling probability using the stationary phase approximation. We concentrate on the case of strong nonlinearity with $C/V \gg 1$, where there is a near-unity tunneling probability to the upper adiabatic level even in the adiabatic limit. This probability can only increase when the sweep rate is finite. We thus expect the amplitude b in the Schrödinger equation (5.34) to remain small and for $|a| \sim 1$ to always hold, and a perturbation treatment of the problem becomes adequate. We begin with the variable transformation,

$$a = a' e^{-i \int_0^t [\frac{\gamma}{2} + \frac{C}{2}(|b|^2 - |a|^2)]dt}, \tag{5.54}$$

$$b = b' e^{i \int_0^t [\frac{\gamma}{2} + \frac{C}{2}(|b|^2 - |a|^2)]dt}. \tag{5.55}$$

As a result, the diagonal terms in Hamiltonian are transformed away, and we have

$$b' = \frac{V}{2i} \int_{-\infty}^{t} dt e^{-i \int_0^t [\gamma + C(|b|^2 - |a|^2)]dt}. \tag{5.56}$$

We need to evaluate the above integral self-consistently. Because of the large C, the nonlinear term in the exponent generally imparts a rapid phase oscillation, which makes the integral small. The dominant contribution comes from the stationary point t_0 of the phase, around which we have

$$-\gamma + C(1 - 2|b|^2) = -\bar{\alpha}(t - t_0), \tag{5.57}$$

with

$$\bar{\alpha} = \alpha + 2C \left[\frac{d}{dt} |b|^2 \right]_{t_0}. \tag{5.58}$$

We thus have

$$|b|^2 = \left(\frac{V}{2} \right)^2 \left| \int_{-\infty}^{t} dt e^{-\frac{i}{2} \bar{\alpha} (t - t_0)^2} \right|^2. \tag{5.59}$$

We can differentiate this expression and evaluate its result at time t_0, obtaining several standard Fresnel integrals with the result $[(d/dt)|b|^2]_{t_0} = (V/2)^2 \sqrt{\pi / \bar{\alpha}}$. Combining this result with the relation (5.58), we obtain a closed equation for $\bar{\alpha}$,

$$\bar{\alpha} = \alpha + 2C \left(\frac{V}{2} \right)^2 \sqrt{\frac{\pi}{\bar{\alpha}}}. \tag{5.60}$$

The nonadiabatic transition probability Γ is given by

$$\Gamma = 1 - |b|^2_{+\infty} = 1 - \left(\frac{V}{2} \right)^2 \left| \int_{-\infty}^{+\infty} dt e^{-\frac{i}{2} \bar{\alpha} (t - t_0)^2} \right|^2 = 1 - \frac{\pi V^2}{2\bar{\alpha}}. \tag{5.61}$$

This result yields a closed equation for Γ [3]:

$$\frac{1}{1 - \Gamma} = \frac{1}{P} + \frac{\sqrt{2}}{\pi} \frac{C}{V} \sqrt{1 - \gamma}, \tag{5.62}$$

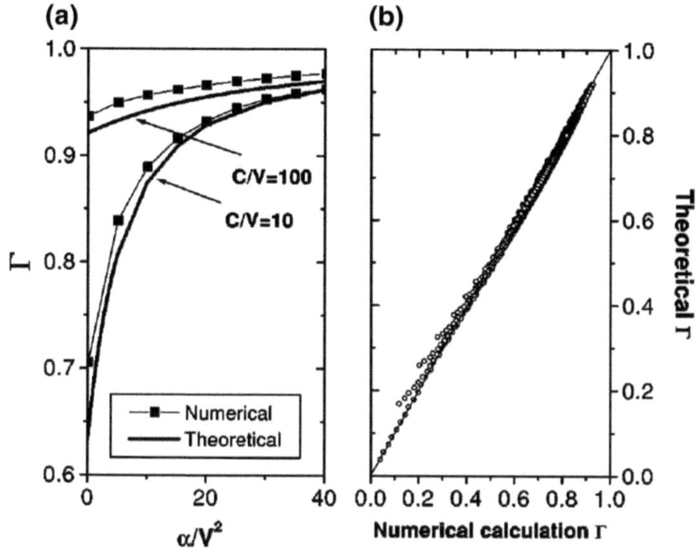

Fig. 5.7 Comparison between our analytic results and the numerical integration of the Schrödinger equation (5.34)

where $P = \pi V^2/(2\alpha)$. In the adiabatic limit, i.e., $1/P = 0$, we find that $\Gamma = 1 - 1.7(V/C)^{2/3}$; in the sudden limit, $1/P \to \infty$, we have $\Gamma = 1 - P$, which is exact. In Fig. 5.7a, we compare the above analytical results with that obtained from directly solving the Schrödinger equation (5.34) and observe that the results are consistent.

The above deduction is made for the case of strong nonlinearity; however, the corresponding result can be extended to a wide range of parameters if we take the quantity P as $1 - \Gamma_{lz} = 1 - e^{-\pi V^2/(2\alpha)}$, representing linear Landau-Zener tunneling. Then, the above equation indicates that the nonlinear tunneling probability is a function of both the linear Landau-Zener tunneling and the ratio between the nonlinear parameter and the energy gap. This relation is confirmed by our numerical calculations. We have calculated the nonlinear tunneling probability using (5.62) with 2500 pairs of Γ_{lz} and C/V, randomly distributed in the ranges $(0.05, 0.95)$ and $(1, 20)$, respectively. These results are compared with the tunneling probabilities obtained by directly integrating the Schrödinger equation (5.34) in Fig. 5.7b, where a very good agreement is observed.

5.2.2 Three-Level System

We consider the simplest multilevel system—the three-level system—to investigate its complex tunneling dynamics in the presence of nonlinearity. Because quantum transitions may occur between several levels simultaneously, the Landau-Zener tunneling in the nonlinear three-level model shows many striking properties distinct from those of the two-level case. We consider the following dimensionless Schrödinger equation:

$$i\frac{d}{dt}\begin{pmatrix} a_1 \\ a_2 \\ a_3 \end{pmatrix} = H\begin{pmatrix} a_1 \\ a_2 \\ a_3 \end{pmatrix}, \tag{5.63}$$

with the Hamiltonian given by

$$H = \begin{pmatrix} \frac{\gamma}{2} + \frac{c}{4}|a_1|^2 & -\frac{v}{2} & 0 \\ -\frac{v}{2} & \frac{c}{4}|a_2|^2 & -\frac{v}{2} \\ 0 & -\frac{v}{2} & -\frac{\gamma}{2} + \frac{c}{4}|a_3|^2 \end{pmatrix}, \tag{5.64}$$

where v is the coupling constant between the neighboring levels; c is the nonlinear parameter; the energy bias γ is assumed to be perturbed by a linear external field, i.e., $\gamma = \alpha t$, where α is the sweep rate; a_1, a_2, and a_3 are the probability amplitudes in each level; and the total probability $|a_1|^2 + |a_2|^2 + |a_3|^2$ is conserved and set to be unity.

When the nonlinear parameter vanishes, our model reduces to the linear case, and the adiabatic energy levels $\varepsilon(\gamma) = 0, \pm\frac{1}{2}\sqrt{\gamma^2 + 2v^2}$ (Fig. 5.8a), derived by diagonalizing the Hamiltonian (5.64). The tunneling probability Γ_{nm} $(n, m = 1, 2, 3)$ is defined as the occupation probability on the mth level at $\gamma \to +\infty$ for the state initially on the nth level at $\gamma \to -\infty$. For the linear case, the above system is solvable analytically, and the tunneling probabilities can be explicitly expressed as [34]

$$\Gamma_{11} = \left[1 - \exp\left(-\frac{\pi v^2}{2\alpha}\right)\right]^2, \tag{5.65}$$

$$\Gamma_{12} = 2\exp(-\frac{\pi v^2}{2\alpha})\left[1 - \exp\left(-\frac{\pi v^2}{2\alpha}\right)\right], \tag{5.66}$$

$$\Gamma_{13} = \exp(-\frac{\pi v^2}{\alpha}), \tag{5.67}$$

$$\Gamma_{22} = \left[1 - 2\exp\left(-\frac{\pi v^2}{2\alpha}\right)\right]^2. \tag{5.68}$$

The others are $\Gamma_{21} = \Gamma_{23} = \Gamma_{32} = \Gamma_{12}$, $\Gamma_{31} = \Gamma_{13}$, and $\Gamma_{33} = \Gamma_{11}$ due to the symmetry of the levels.

With the presence of the nonlinear terms, we wish to know how the tunneling dynamics in the above system are affected. In our discussions, the coupling parameter is set to unity, i.e., $v = 1$. Therefore, the weak nonlinearity case and strong nonlinearity case correspond to $c \ll 1$ and $c \gg 1$, respectively.

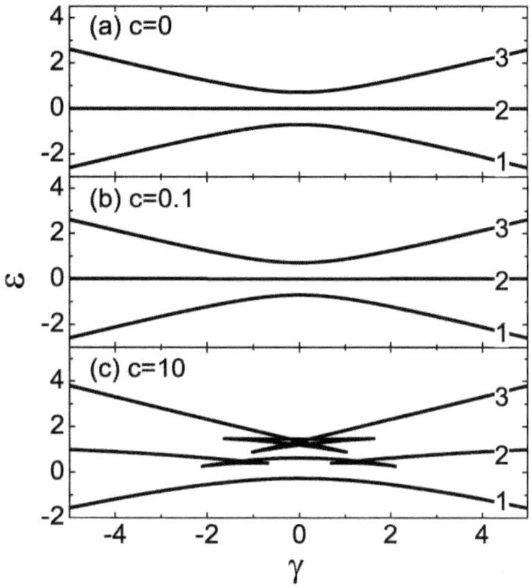

Fig. 5.8 Adiabatic energy levels at $v = 1.0$: **a** linear case, **b** weak nonlinearity case of $c = 0.1$, **c** strong nonlinearity case of $c = 10.0$

Similar to the linear case, we need to analyze the adiabatic levels of the nonlinear model first. With $a_1 = \sqrt{s_1}e^{i\theta_{a_1}}$, $a_2 = \sqrt{1-s_1-s_2}e^{i\theta_{a_2}}$, and $a_3 = \sqrt{s_2}e^{i\theta_{a_3}}$, we introduce the relative phases $\theta_1 = \theta_{a_1} - \theta_{a_2}$ and $\theta_2 = \theta_{a_3} - \theta_{a_2}$. In terms of s_1, θ_1, s_2, and θ_2, the nonlinear three-level system is cast into a classical Hamiltonian system,

$$H_e = \left(\frac{\gamma}{2} + \frac{c}{8}s_1\right) s_1 + \frac{c}{8}(1 - s_1 - s_2)^2 + \left(-\frac{\gamma}{2} + \frac{c}{8}s_2\right) s_2$$
$$- v\sqrt{(1-s_1-s_2)s_1}\cos\theta_1 - v\sqrt{(1-s_1-s_2)s_2}\cos\theta_2. \tag{5.69}$$

s_1, θ_1 and s_2, θ_2 are two pairs of canonically conjugate variables of the classical Hamiltonian system, governed by the following differential equations:

$$\dot{s_1} = -v\sqrt{(1-s_1-s_2)s_1}\sin\theta_1, \tag{5.70}$$

$$\dot{\theta_1} = \frac{\gamma}{2} - \frac{c}{4}(1-2s_1-s_2) - \frac{1-2s_1-s_2}{2\sqrt{(1-s_1-s_2)s_1}}v\cos\theta_1 + \frac{s_2}{2\sqrt{(1-s_1-s_2)s_2}}v\cos\theta_2, \tag{5.71}$$

$$\dot{s_2} = -v\sqrt{(1-s_1-s_2)s_2}\sin\theta_2, \tag{5.72}$$

$$\dot{\theta_2} = -\frac{\gamma}{2} - \frac{c}{4}(1-s_1-2s_2) + \frac{s_1}{2\sqrt{(1-s_1-s_2)s_1}}v\cos\theta_1 - \frac{1-s_1-2s_2}{2\sqrt{(1-s_1-s_2)s_2}}v\cos\theta_2. \tag{5.73}$$

The fixed points of the nonlinear classical Hamiltonian correspond to the eigenstates of the nonlinear three-level system. By setting $\dot{s_1} = \dot{s_2} = \dot{\theta_1} = \dot{\theta_2} = 0$

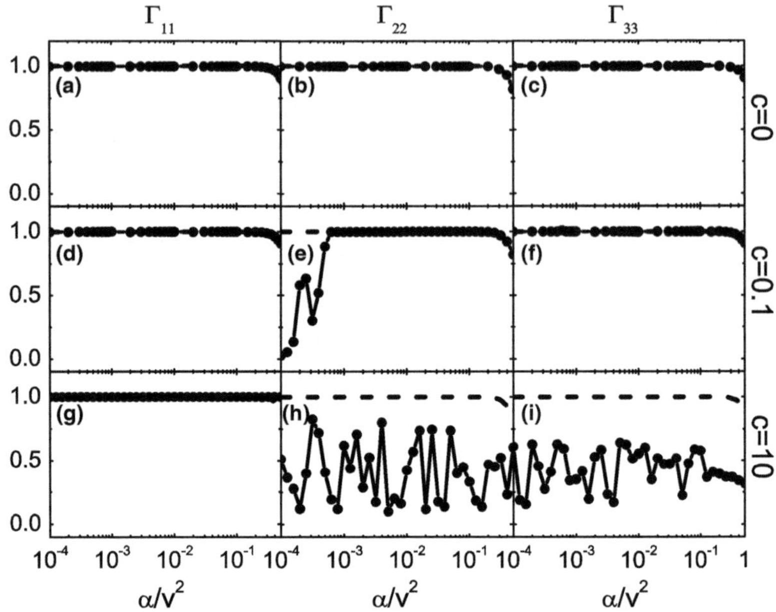

Fig. 5.9 The tunneling probabilities Γ_{11}, Γ_{22}, and Γ_{33} (full circles) as functions of α for different nonlinear parameters at $v = 1.0$. The dashed lines represent the results from the linear Landau-Zener model for comparison

in Eqs. (5.70)–(5.73), the eigenstates of the system are obtained. Accordingly, the eigenenergy is obtained via $\varepsilon = H_e$, i.e., the energy levels are obtained, as shown in Fig. 5.8.

For weak nonlinearity, the levels' structure is similar to its linear counterpart (Fig. 5.8b). For strong nonlinearity (Fig. 5.8c), in the middle level, a double-loop topological structure emerges; in the upper level, a butterfly structure appears. Because of these topological distortions on the energy levels, we expect that the tunneling dynamics will dramatically change.

In the adiabatic limit, the characters of the tunneling probabilities are expected to be entirely determined by the topology of the energy levels and the eigenstates' properties (corresponding to the stability of the fixed points in the classical Hamiltonian system), according to the adiabatic theorem [4]. Thus, we expect that for the weak-nonlinearity case, an initial state starting from any level (upper, middle or lower) follows the levels and evolves adiabatically; as a result, no quantum transition between levels occurs. For the strong nonlinearity case, an initial state from the lower level is expected to evolve adiabatically and remain on the ground state, leading to zero adiabatic tunneling probability, whereas the state initially corresponding to the middle or upper level cannot move smoothly from the left side to the right side due to the topological change of the level. A transition to other levels occurs at the tip of

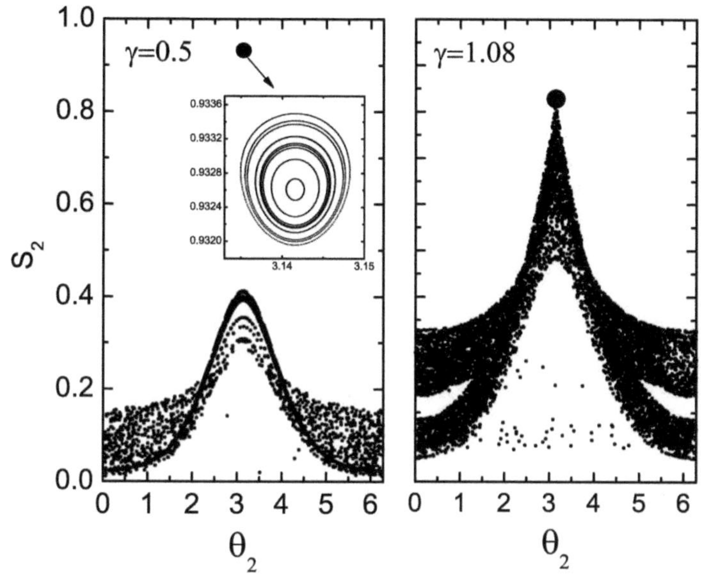

Fig. 5.10 Poincaré section of the trajectories for $c = 10$ before and after the tip of the butterfly structure of the upper level in Fig. 5.8c

the loop or butterfly. Consequently, the adiabatic tunneling probability is expected to be nonzero.

However, this picture is only partly corroborated by our direct solution of the Schrödinger equation using a 4–5th-order Runge-Kutta adaptive-step algorithm, as shown in Fig. 5.9.

On the one hand, as we expect, Fig. 5.9 clearly shows—for the strong nonlinearity case—no tunneling for the state from the lower level; however, substantial adiabatic tunneling is observed for the states from the upper two levels. In particular, we find that the tunneling probability as a function of the sweep rate shows an irregular oscillation. This oscillation was also reported by Graefe et al. [35]. We associate this irregularity with the chaotic state. To demonstrate this state, we plot in Fig. 5.10 the Poincaré section of the trajectories for $c = 10$ before and after the tip of the butterfly structure of the upper level in Fig. 5.8c. Before the tip, the eigenstate corresponds to the fixed point surrounded by the quasiperiodic orbit; therefore, this state is stable. As the state evolves to the right-hand tip of the butterfly, it makes contact with the chaotic sea, and the states subsequently become chaotic. The characteristics of the chaos are sensitive to the input parameters; therefore, the chaotic state is responsible for the irregular tunneling probability revealed by Fig. 5.9h and i.

On the other hand, Fig. 5.9 also shows that for the case of weak nonlinearity, even though the adiabatic levels keeps the same topological structure as in the linear case, a nonzero tunneling probability remains for the state starting from the middle level. The tunneling also shows some kind of irregularity. This phenomenon contrasts

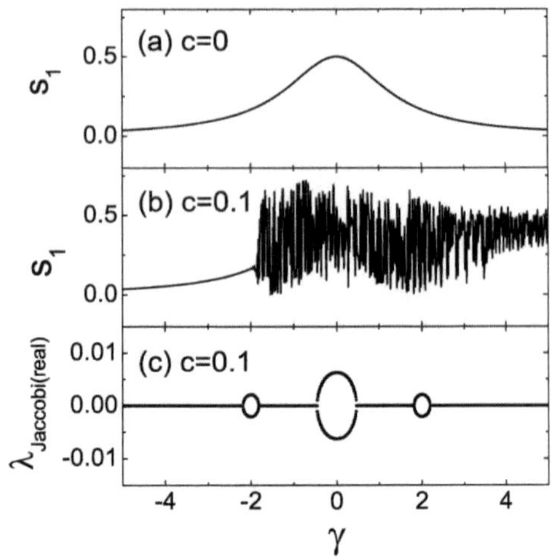

Fig. 5.11 Variation of s_1 with γ when the eigenstate $(0, 1, 0)^T$ evolves adiabatically at $v = 1.0$, $\alpha = 0.0001$. **a** Linear case. **b** Nonlinear case at $c = 0.1$. **c** Real parts of the eigenvalues of H_J

with our naive conjecture developed from observing the topological structure of the adiabatic levels.

To explain this unusual phenomenon, we conduct a detailed analysis of the properties of the fixed points of the classical system Hamiltonian (5.69), corresponding to the eigenstates of the middle level.

We plot the quantity s_1 as a function of γ in Fig. 5.11a and b and observe that the adiabatic evolution of the eigenstate breaks down at approximately $\gamma = -2$ due to the nonlinearity (Fig. 5.11b). This adiabaticity breakage is caused by the change in the properties of the fixed point corresponding to the eigenstate of the middle level. This aspect is revealed by investigating the Hamiltonian-Jacobi matrix obtained by linearizing the nonlinear Eqs. (5.70)–(5.73) at fixed points:

$$H_J = \begin{pmatrix} -\frac{\partial^2 H_e}{\partial s_1 \partial \theta_1} & -\frac{\partial^2 H_e}{\partial^2 \theta_1} & -\frac{\partial^2 H_e}{\partial s_2 \partial \theta_1} & -\frac{\partial^2 H_e}{\partial \theta_2 \partial \theta_1} \\ \frac{\partial^2 H_e}{\partial^2 s_1} & \frac{\partial^2 H_e}{\partial \theta_1 \partial s_1} & \frac{\partial^2 H_e}{\partial s_2 \partial s_1} & \frac{\partial^2 H_e}{\partial \theta_2 \partial s_1} \\ -\frac{\partial^2 H_e}{\partial s_1 \partial \theta_2} & -\frac{\partial^2 H_e}{\partial \theta_1 \partial \theta_2} & -\frac{\partial^2 H_e}{\partial s_2 \partial \theta_2} & -\frac{\partial^2 H_e}{\partial^2 \theta_2} \\ \frac{\partial^2 H_e}{\partial s_1 \partial s_2} & \frac{\partial^2 H_e}{\partial \theta_1 \partial s_2} & \frac{\partial^2 H_e}{\partial^2 s_2} & \frac{\partial^2 H_e}{\partial \theta_2 \partial s_2} \end{pmatrix}. \tag{5.74}$$

We obtain the eigenvalues of H_J for different γ and plot our results in Fig. 5.11c. These eigenvalues can be real, complex or purely imaginary. Only purely imaginary eigenvalues correspond to the stable fixed point; others indicate unstable ones. Figure 5.11c shows that the eigenvalues are complex numbers (i.e., their real parts are not zero) around $\gamma = 0, \pm 2$. The corresponding fixed points are unstable. For other regions, the eigenvalues of H_J are purely imaginary. Therefore, even though the topological structure does not change in terms of the level structures, the instability of the fixed point corresponding to the middle level leads to the breakdown of the adiabaticity, manifested in the irregular nonzero tunneling probability revealed by Fig. 5.9e in the adiabatic limit.

The above instability mechanism occurs for any small nonlinear perturbation. We present selected analytic deductions as follows. Note that the fixed points of Eqs. (5.70)–(5.73) can be accurately calculated if $c = 0$: $s_1^0 = s_2^0 = \frac{1}{2+\gamma^2}$, $\theta_1^0 = 0$ and $\theta_2^0 = \pi$ for $\gamma > 0$, and $\theta_1^0 = \pi$ and $\theta_2^0 = 0$ for $\gamma < 0$. By employing perturbation theory using c as the small parameter, we obtain the fixed points for small c: $s_1^0 = \frac{1}{2+\gamma^2} - \frac{(1-\gamma^2)^2}{4(2+\gamma^2)} c\gamma$, $s_2^0 = \frac{1}{2+\gamma^2} + \frac{(1-\gamma^2)^2}{4(2+\gamma^2)} c\gamma$, $\theta_1^0 = 0$, and $\theta_2^0 = \pi$ for the nonlinear case. Substituting these points into Eq. (5.74), we can obtain the eigenvalues of H_J by solving the following quartic equation:

$$(64 + 1280\gamma^4)x^4 + (64 + c^2 + 1344\gamma^2)x^2 + (16 + c^2 + 352\gamma^2) = 0. \tag{5.75}$$

The useful quadratic discriminant is $\Delta = 4096\gamma^4 - 2432c^2\gamma^2 + (c^4 - 128c^2)$. In the linear case, $c = 0$, $\Delta = 4096\gamma^4$ is always larger than zero, which means that the solutions for x are purely imaginary; thus, the fixed points are stable. For small c, $\lim_{\gamma\to 0}\Delta < 0$, the real part of the solutions $\sim c/16$, while the imaginary part $\sim\sqrt{2}/2$. As a result, the fixed point corresponding to the middle level becomes unstable around $\gamma = 0$ for any small nonlinearity, implying the breakdown of the adiabatic evolution of states for the middle level.

5.2.3 *Spatially Magnetic Modulated Trap*

In this subsection, we study the nonlinear Landau-Zener tunneling of a Bose-Einstein condensate in a spatially magnetic modulated trap as shown in Fig. 5.12, which is a single well with a spatially modulated scattering length, actually called a pseudo double-well potential [36]. In regions (a) and (c), the s-wave scattering length $a_s = 0$; in region (b), however, $a_s \neq 0$, which allows tuning with a magnetic-field Feshbach resonance. The wave function satisfies the following nonlinear equation:

$$i\frac{\partial\Psi}{\partial t} = -\frac{d^2}{dx^2}\Psi + V(x)\Psi + a_s(x)|\Psi|^2\Psi. \tag{5.76}$$

For the present pseudo double-well potential, the wave function $\Psi(x,t)$ is described by a superposition of the two modes of different symmetry, i.e., symmetric and antisymmetric [37]:

$$\Psi(x,t) = c_1(t)\phi_L(x) + c_2(t)\phi_R(x). \tag{5.77}$$

Here, $\phi_{L,R}(x) = [\phi_1(x) \pm \phi_2(x)]/\sqrt{2}$ with $\phi_{1,2}(x)$ are the symmetric and antisymmetric functions, respectively [37, 38]. In Eq. (5.77), c_1 and c_2 are the probability amplitudes of atoms in each of the effective two wells. The leading goal of this work is to investigate the effect of the scattering length a_s, which can be widely tuned with a magnetic-field Feshbach resonance, on the energy landscape and corresponding Landau-Zener tunneling, so we take the potential energy $V(x) = 0$ for convenience.

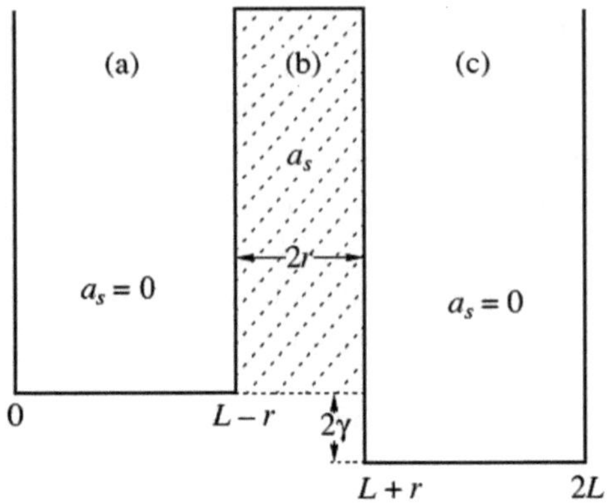

Fig. 5.12 Schematic sketch of our model. In regions (**a**) and (**c**), the s-wave scattering length $a_s = 0$; however, in region (**b**), $a_s \neq 0$, which allows tuning with a magnetic-field Feshbach resonance. We call this model a pseudo double-well potential, where 2γ is the level separation of the zero-point energy. In this work, we focus on the case of repulsive interaction between atoms, i.e., $a_s > 0$

The effective double-well system is described by a dimensionless Schrödinger equation,

$$i\frac{d}{dt}\begin{pmatrix} c_1 \\ c_2 \end{pmatrix} = H(\gamma)\begin{pmatrix} c_1 \\ c_2 \end{pmatrix}, \tag{5.78}$$

with the Hamiltonian given by

$$H(\gamma) = \begin{pmatrix} H_{11} & H_{12} \\ H_{21} & H_{22} \end{pmatrix}, \tag{5.79}$$

where

$$H_{11} = \gamma + c|c_1|^2 + \eta\Delta + \omega|c_2|^2, \tag{5.80}$$

$$H_{12} = H_{21} = -v + \eta + \omega\Delta, \tag{5.81}$$

$$H_{22} = -\gamma + c|c_2|^2 + \eta\Delta + \omega|c_1|^2, \tag{5.82}$$

with

$$\Delta = c_1^* c_2 + c_1 c_2^*. \tag{5.83}$$

The total probability $|c_1|^2 + |c_2|^2$ is conserved and is set to unity. In the above derivation, 2γ is the level separation, v is the number-independent coupling constant between the two levels, η and ω describe the coupling energy dependence on the level population difference due to the exchange energy between atoms of different states, and c is the nonlinear parameter representing the interactions between atoms belonging to the same state [3]. For the symmetric case $\gamma = 0$, the expressions of v, c, η, and ω are as follows:

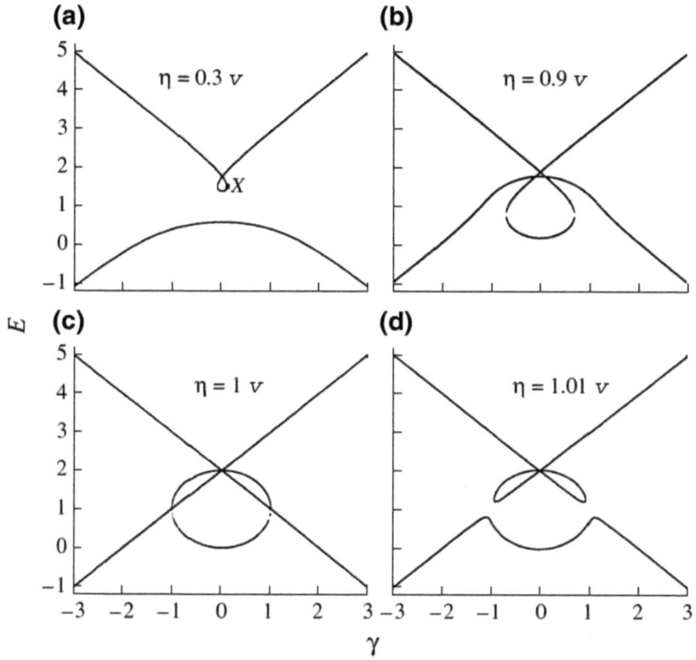

Fig. 5.13 Adiabatic energy levels for different η. Here, we set $v = 1, \omega = 0$, and interaction strength $c = 2v$

$$v = \int_0^{2L} \phi_L \frac{d^2\phi_R}{dx^2} dx = \int_0^{2L} \phi_R \frac{d^2\phi_L}{dx^2} dx, \tag{5.84}$$

$$c = a_s \int_{L-r}^{L+r} \phi_L^4 dx = a_s \int_{L-r}^{L+r} \phi_R^4 dx, \tag{5.85}$$

$$\eta = a_s \int_{L-r}^{L+r} \phi_L \phi_R^3 dx = a_s \int_{L-r}^{L+r} \phi_R \phi_L^3 dx, \tag{5.86}$$

$$\omega = a_s \int_{L-r}^{L+r} \phi_L^2 \phi_R^2 dx. \tag{5.87}$$

Relative to previous models, the introduction of the parameters η and ω makes the energy landscape more complex. With this effective two-level model, the energy as a function of the level bias has different structures corresponding to different parameters, as shown in Figs. 5.13 and 5.14. The consequence of the loop structure shown in Fig. 5.13a, as discussed in [3], is that a quantum state moves along the upper level by changing γ so slowly that little tunneling to the lower level is generated. The state gains energy until reaching the singular point X in Fig. 5.13a, and no options

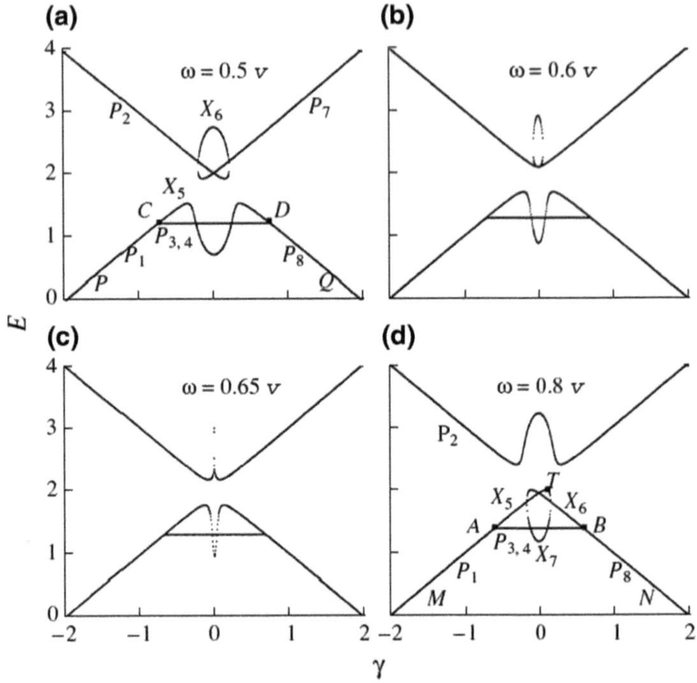

Fig. 5.14 Adiabatic energy levels for different ω, with $v = 1$, $c = 2v$, and $\eta = 1.01v$. In **a** and **d**, P_i points correspond to stable elliptic points, and X_i points are unstable saddle points

remain for further movement except to jump to the lower and upper levels. As a result, the nonlinear Landau-Zener tunneling probability is not zero even in the adiabatic limit $\alpha \to 0$. For another typical nonlinear case, as shown in Fig. 5.14, the lower level develops a horizontal structure. In this structure (shown in Fig. 5.13d, for example), notably, the quantum state starting from the lower level can never move to T because it remains in a stable state after arriving at A.

To avoid the challenges associated with nonlinearity, we place our analysis on a firm mathematical basis by studying an equivalent classical Hamiltonian. In this transform, the eigenstates become fixed points in the classical problem, and the nonzero adiabatic tunneling probability is interpreted as the result of collision between fixed points.

With $c_1 = \sqrt{s_1}e^{i\theta_1}$ and $c_2 = \sqrt{s_2}e^{i\theta_2}$, we introduce the population difference

$$n = s_2 - s_1 \tag{5.88}$$

and the relative phase

$$\theta = \theta_2 - \theta_1. \tag{5.89}$$

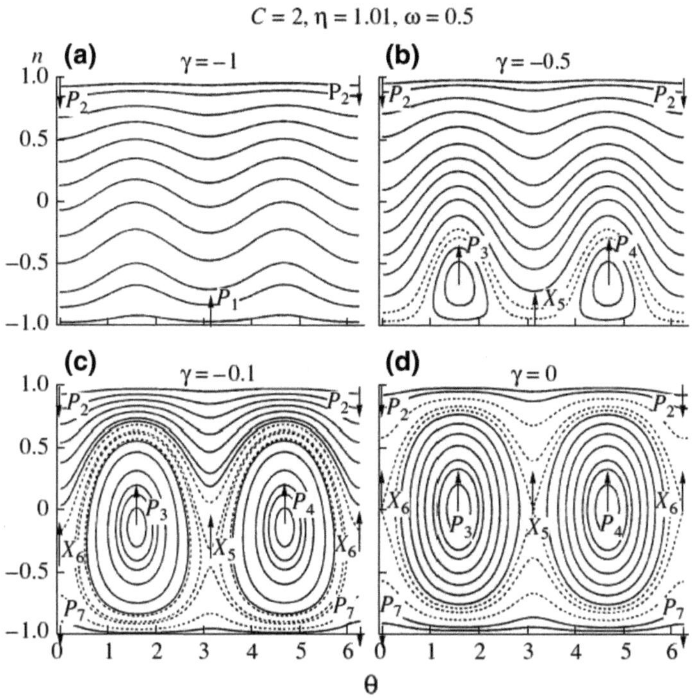

Fig. 5.15 Evolution of the phase-space motions of the Hamilton H_c system as γ changes adiabatically for the same parameters as in Fig. 5.14a. Points X_i are saddle points, and points P_i are elliptic points. The arrows indicate the shifting direction of the fixed point as γ increases

In terms of n and θ, the effective nonlinear two-level system is cast as a classical Hamiltonian system,

$$H_c = 2\gamma n - \frac{1}{2}(c - 2\omega)n^2 + 2(v - \eta)\sqrt{1 - n^2}\cos\theta - \frac{1}{2}\omega(1 - n^2)\cos 2\theta, \tag{5.90}$$

where n and θ are a pair of canonically conjugate variables of the classical Hamiltonian system. The fixed points of the nonlinear classical Hamiltonian correspond to the eigenstates of the quantum system and are given by the following equations:

$$\dot{n} = 2(v - \eta)\sqrt{1 - n^2}\sin\theta - \omega(1 - n^2)\sin 2\theta, \tag{5.91}$$

$$\dot{\theta} = 2\gamma - (c - 2\omega)n - \frac{2(v - \eta)n}{\sqrt{1 - n^2}}\cos\theta + \omega n \cos 2\theta. \tag{5.92}$$

By solving Eqs. (5.91) and (5.92), the eigenstates of the system are obtained. We restrict our interest to the region enclosed by $-1 \leq n \leq 1$ and $0 \leq \theta \leq 2\pi$. Many interesting results emerge during the study of this set of equations.

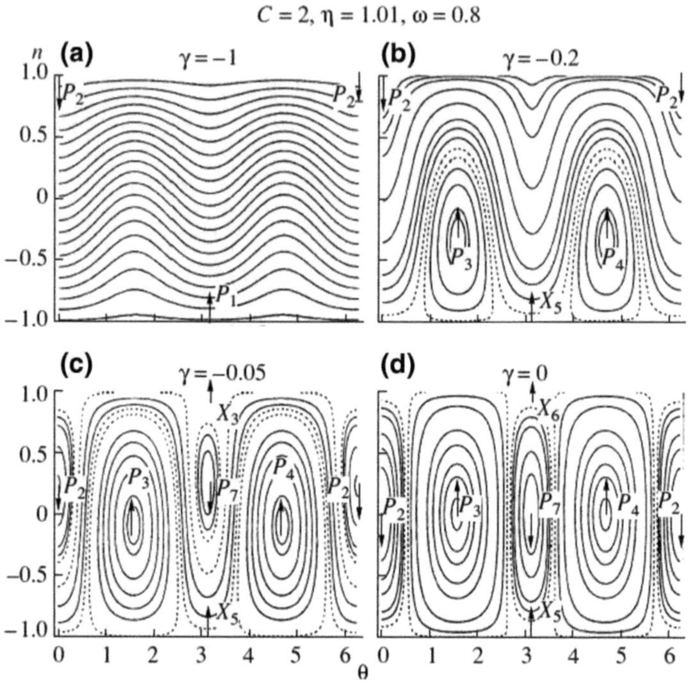

Fig. 5.16 Evolution of the phase-space motions of the Hamilton H_c system as γ changes adiabatically for the same parameters as in Fig. 5.14d. Points X_i are saddle points, and points P_i are elliptic points. The arrows indicate the shifting direction of the fixed point as γ increases

Deriving the analytical expressions of the fixed points is challenging; however, we can readily obtain the equivalent results numerically. The numerical results are shown in Figs. 5.15 and 5.16, corresponding to Fig. 5.14a and b, respectively. Points X_i are saddle points, and points P_i are elliptic points, each being surrounded by a closed (elliptic) orbit. In Fig. 5.15, the collision between P_2 and X_6 leads to nonzero adiabatic tunneling from the upper level to the lower level, which is determined by the ultimate fate of the fixed point P_2. As the level bias changes from $\gamma = -\infty$ to $\gamma = +\infty$, P_2 in Fig. 5.16 moves smoothly along the line $\theta = 0$ from the top ($s = 1$) to the bottom ($s = -1$), corresponding to the upper level in Fig. 5.14d. The original fixed point P_1 moves smoothly up to C [in Fig. 5.14a] or A [in Fig. 5.14d], where two more fixed points appear at $\theta = \pi/2$ and $3\pi/2$ [P_3 and P_4 in Figs. 5.15b–d or 5.16b–d, corresponding to the line CD in Fig. 5.14a or the line AB in Fig. 5.14d], all being elliptic. The state along CD or AB is stable, i.e., any deviation of the initial condition of the system cannot cause the trajectory in the phase space to move away from this line. The state continues to move along the line as γ increases, and no tunneling occurs.

We now turn to Landau-Zener tunneling in the pseudo nonlinear two-level system. In the following discussion, we focus on the case of repulsive interaction between atoms, i.e., $c > 0$. As in the linear Zener model, we take v to be independent of time, and we wish to examine the nonlinear Landau-Zener tunneling, i.e., how the system evolves when the level bias γ changes with time as $\gamma = \alpha t$. We take α to be the sweep rate. In this work, we focus on the adiabatic limit, that is, the case in which the sweep rate α tends to zero. In the adiabatic limit, the characters of the tunneling probabilities are determined entirely by the topology of the energy levels and the eigenstates properties, which correspond to the stability of the fixed points in the classical Hamiltonian system [3, 4]. In the study of the tunneling probability for the nonlinear Landau-Zener model, we are interested in the long-time solution of Eq. (5.78) provided that the system is prepared at $t \to -\infty$ in the high- or low-energy stationary state. For example, we choose the low-energy c_1, i.e., $c_1^0(t \to -\infty) = 1$, and the tunneling probability is thus $P_T = |c_1(t \to \infty)|^2$. In the linear case, the tunneling probability is independent of whether the system starts in the low- or high-energy state; an initial state starting from any level (upper or lower) follows the level and evolves adiabatically. As a result, no quantum transition between levels occurs. For the nonlinear case, the energy levels (upper and/or lower) develop a loop structure corresponding to special parameters. An initial state from the level that is a smooth curve is expected to evolve adiabatically while remaining in the ground state, leading to zero adiabatic tunneling probability, whereas for the state initially from the level developing a loop, due to the topological change of the level, the state cannot move smoothly from the left side to the right side. A transition to another level occurs at the tip of the loops. Consequently, the adiabatic tunneling probability is expected to be nonzero.

The breakdown of adiabatic evolution is confirmed by numerical calculation of the tunneling probability directly from Eq. (5.78) for two initial states, $(c_1^0, c_2^0) = (1, 0)$ and $(c_1^0, c_2^0) = (0, 1)$, corresponding to the initial state starting from the lower and upper levels, respectively. The numerical results of the probability for different parameters are shown in Figs. 5.17, 5.18 and 5.19. The small fluctuations appearing in the curves occur because the numerical simulation must start at some finite time instead of the ideal limit of $t = -\infty$.

As mentioned above, in the linear case $c = 0$, the instantaneous eigenvalue independence of $\gamma = \alpha t$ (the so-called adiabatic levels) form a consecutive avoided crossing; consequently, the transition probability for the linear case is independent of other parameters. This condition no longer holds in the nonlinear case since the influence of the nonlinearity, e.g., the emergence and structure of the loops, depends on v, η, ω, and γ. This relationship is confirmed by our numerical results shown in Fig. 5.17, where $P_T(\alpha)$ is plotted for different values of ω for the same parameters as in Fig. 5.14. For the state initially from the lower levels $(c_1^0, c_2^0) = (1, 0)$ as shown in Fig. 5.17a, there appears a damped oscillating tunneling probability that still tends to zero in the adiabatic limit ($P_T \to 0$ for $\alpha \to 0$) for small values of the sweep rate α with decreasing α. These results can be understood through the following argument. As shown in Fig. 5.14, the lower energy level develops a horizontal structure. For example, in Fig. 5.14d, the quantum state on the horizontal segment AB of the lower level is a stable state, while the state corresponding to AT is unstable. The quantum state starting from the lower level can never move to T because once it arrives at A, it remains at

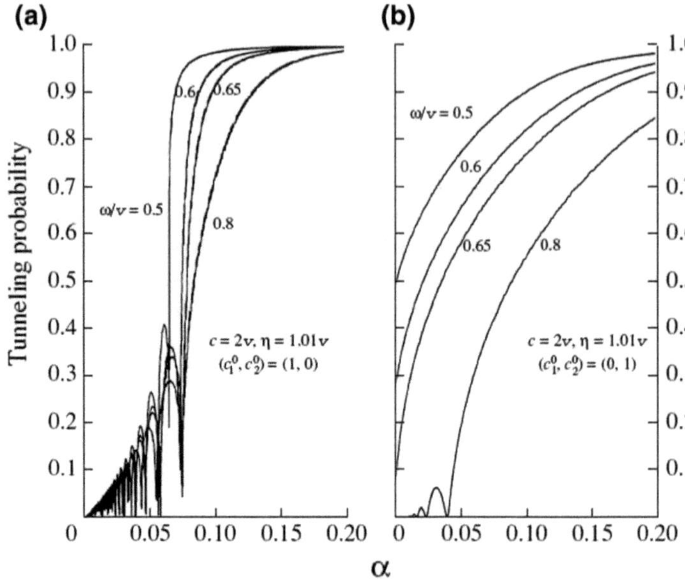

Fig. 5.17 Tunneling probability as a function of the sweep rate α for different values of ω for an initial state starting from the lower level (**a**) and upper level (**b**), with $c = 2v$ and $\eta = 1.01v$

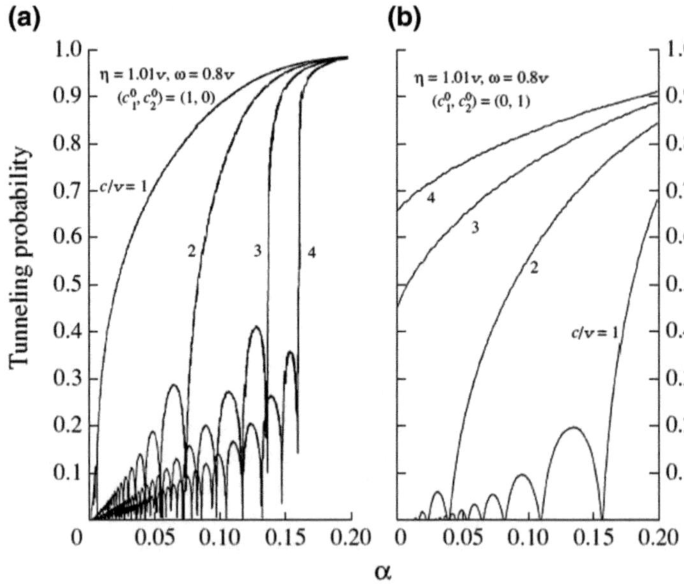

Fig. 5.18 Tunneling probability for different nonlinear parameters c, with $\eta = 1.01v$ and $\omega = 0.8v$

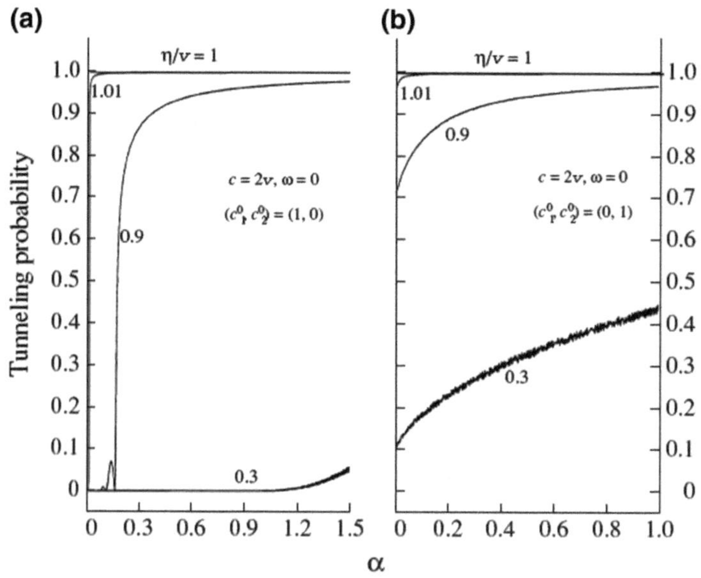

Fig. 5.19 Tunneling probability for the same parameters as in Fig. 5.13

a stable state, i.e., the state continues moving along line AB as γ increases and then moves smoothly from the left side to the right side along $M - A - B - N$, leading to zero adiabatic tunneling probability. The numerical results for an initial state starting from the upper levels are shown in Fig. 5.17b; here, the tunneling increases with decreasing ω. Most strikingly, the tunneling probability for $\omega \leq 0.65v$ is not zero in the adiabatic limit $\alpha \to 0$ for the appearance of loops shown in Fig. 5.14, while it goes to zero for $\omega \leq 0.65v$ [$\omega = 0.8v$ in Fig. 5.17b], as no loops have occurred yet. At the critical value $\omega = 0.65v$, the transition probability vanishes with a nonzero slope.

Figure 5.18 plots the dependence of the probability on α for different values of the nonlinear parameter c, with $\eta = 1.01v$ and $\omega = 0.8v$. As shown in Fig. 5.18a, for the state initially starting from the lower levels, a novel feature is the appearance of a pronounced damped oscillation of the tunneling probability with decreasing α for small values of α. A stronger interaction strength corresponds to a broader oscillating region. Notably, the tunneling probabilities for $c \geq v$ are zero in the adiabatic limit $\alpha \to 0$, while the probability is assumably not zero for one critical value of $c < v$. The numerical results of tunneling for an initial state starting from the upper level increases with increasing interaction strength. For weak nonlinearities, e.g., $c = 1v$ and $2v$ in Fig. 5.18b, the probability still tends to zero in the adiabatic limit, i.e., $P_T \to 0$ for $\alpha \to 0$, as no loops have occurred yet. This condition is no longer true after the appearance of the loops ($c = 3v$ and $4v$) such that $P_T(\alpha \to 0) > 0$.

We calculate the probability for the same parameters as in Fig. 5.13. The numerical results are shown in Fig. 5.19. For $\omega = 0$ and $\eta = 1v$, the off-diagonal component of the Hamiltonian is zero; thus, an initial state starting from the lower or upper levels cannot follow the level and inevitably transitions to another level, i.e., the tunneling probability $P_T = 1$. For other values of η, the lower energy levels do not develop loop structures, so the probability for the initial state tunneling from a low energy is zero. In contrast, for an initial state from the upper level, the adiabatic tunneling probabilities are not zero because of the loop structures. For $\eta < 1v$, the energy gap decreases with increasing η, so the probability is large for large values of η. The results for $\eta > 1v$ are reversed.

5.3 Nonlinear Rosen-Zener Tunneling

The Rosen-Zener model was first proposed to study the spin-flip of two-level (hyperfine Zeeman energy level) atoms interacting with a rotating magnetic field by Rosen and Zener to account for the double Stern-Gerlach experiments [39]. In contrast to the well-known Landau-Zener model that depicts the tunneling dynamics between two avoided-crossing energy levels [22], in the Rosen-Zener model, the energy bias between two levels is fixed, and the coupling between two modes is time dependent as described by a rectangular [40], Gaussian [41], exponential [42], or hyperbolic-secant function [39]. This model has attracted much attention not only because it has an exact analytic solution, which provides a way to understand complex multimode systems [43], but also due to its versatile applications in the nonresonant charge exchange of ion-atom collisions [44], laser-induced excitation [45], nuclear magnetic resonance techniques [46], and quantum computation [47], to name only a few.

In the present section, we extend the Rosen-Zener model to the nonlinear case and examine how nonlinearity affects the quantum transition dynamics. We find that the nonlinearity can affect the quantum transition dramatically. At a certain level of nonlinearity, 100% population transfer between two levels is observed and found to be robust over a very wide range of external parameters. On the other hand, the quantum transition can be completely blocked by a strong nonlinearity. In the adiabatic limit, we have derived analytical expressions for the transition probability [48].

The nonlinear two-mode system we consider is described by the following dimensionless Schrödinger equation:

$$i\frac{\partial}{\partial t}\begin{pmatrix} a \\ b \end{pmatrix} = H(t)\begin{pmatrix} a \\ b \end{pmatrix}, \tag{5.93}$$

with the Hamiltonian given by

$$H(t) = \left[\frac{\gamma}{2} + \frac{c}{2}(|b|^2 - |a|^2)\right]\hat{\boldsymbol{\sigma}}_z + \frac{\nu}{2}\hat{\boldsymbol{\sigma}}_x, \tag{5.94}$$

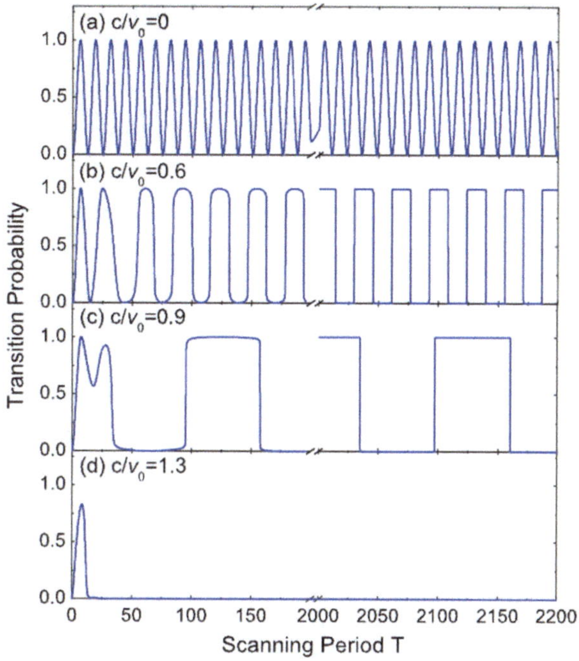

Fig. 5.20 Numerical results for the transition probability versus the scanning period with different nonlinear parameters c/v_0 **b** 0.6, **c** 0.9, and **d** 1.3. For comparison, we also include the results of the linear case in (**a**), which accurately reproduces the results predicted by formula (5.97)

where $\hat{\boldsymbol{\sigma}}_x$ and $\hat{\boldsymbol{\sigma}}_z$ are the Pauli matrices. γ and v are the energy bias and coupling strength between the two modes, respectively. c is the nonlinear parameter describing the interatomic interaction. The total probability $|a|^2 + |b|^2$ is conserved and set to 1 without loss of generality.

In contrast to the nonlinear Landau-Zener model, where the coupling strength remains constant and the energy bias varies linearly in time [2, 3, 22], in this model, the energy bias γ and nonlinear parameter c are set to be constant, whereas the coupling strength v is time dependent and governed by an external pulse field of the form [48]

$$v = \begin{cases} 0, & t < 0, t > T, \\ v_0 \sin^2\left(\frac{\pi t}{T}\right), & t \in [0, T], \end{cases} \tag{5.95}$$

where T is the scanning period of the external field.

In the following discussion, we assume that the quantum state is prepared on one mode initially. With the external field turned on, a quantum transition between different modes emerges. Our focus is the population dynamics in the presence of the external field. The transition probability Γ is defined as the probability of the particle occupying the other mode after the coupling field has been turned off.

We start our analysis with the simplest case, in which both the energy bias and nonlinear parameter vanish ($\gamma = 0$ and $c = 0$). In this case, the Schrödinger equation denoted by (5.93) is solvable analytically. Setting $(a, b) = (1, 0)$ as the initial condition, we readily obtain the probability of the particle populated on the other mode as a function of time,

$$p(t) = |b(t)|^2 = \sin^2\left(v_0 \frac{2\pi t - T\sin\frac{2\pi t}{T}}{8\pi}\right). \tag{5.96}$$

Then, the total transition probability is obtained by substituting $t = T$ into the above equation:

$$\Gamma = p(T) = \sin^2\left(\frac{v_0 T}{4}\right). \tag{5.97}$$

This expression demonstrates a perfect Rabi-like oscillation of the transition probability versus the pulse duration or scanning period T of the external field. The oscillation frequency is proportional to the maximum coupling strength v_0.

For the degenerate case $\gamma = 0$, with the emergence of nonlinearity, the transition dynamics dramatically changes. In this case, (5.93) is no longer analytically solvable. We therefore exploit a 4–5th-order Runge-Kutta algorithm to trace the quantum evolution numerically and calculate the transition probability. In our calculation, we choose the maximum coupling strength v_0 as the energy scale; thus, the weak nonlinearity and strong nonlinearity refer to $c/v_0 \ll 1$ and $c/v_0 \gg 1$, respectively.

Our numerical results are presented in Fig. 5.20, where the transition probability as a function of the scanning period is plotted against the nonlinear parameters that range from weak nonlinearity to strong nonlinearity. Figure 5.20a shows a regular oscillating pattern, consistent with the analytic prediction of formula (5.97). This regular periodic pattern is disrupted with the emergence of nonlinearity. For the weak nonlinear case $c/v_0 < 1$, the periodicity is lost only in the short pulse regime, i.e., $0 < T < 50$ at $c/v_0 = 0.9$. In contrast, in the regime of the large scanning period, a rectangular periodic pattern emerges instead. The period of the rectangular oscillation increases with increasing nonlinearity. This pattern is of particular interest in practice because it represents a complete population transfer between two modes robustly over a wide parameter regime.

For the case of strong nonlinearity, i.e., $c/v_0 > 1$ (see Fig. 5.20d), the quantum transition between two modes is even more strongly affected by the nonlinearity. The oscillation pattern is completely broken, and when the scanning period $T > 15$, the quantum transition is completely blocked.

The above phenomena are intriguing and need detailed consideration. Our further analysis focuses only on the adiabatic limit, which implies that the external field varies slowly relative to the intrinsic motion of the system. The formula (5.97) indicates a period of intrinsic motion that is characterized by $4\pi/v_0$, while the external field is characterized by T. Thus, the adiabatic limit implies that $T \gg 4\pi/v_0$ or $v_0 T \gg 4\pi$. In the following, we derive selected analytical formulas for the transition probability and explain the above phenomena.

According to the adiabatic theory of nonlinear quantum mechanics [4], the characteristics of a quantum transition in the adiabatic limit should be entirely determined by the structure of the energy levels and the properties of the corresponding eigenstates. The eigenstates of the system satisfy

$$\left[\frac{c}{2}(|b|^2 - |a|^2)\hat{\sigma}_z + \frac{v}{2}\hat{\sigma}_x\right]\begin{pmatrix} a \\ b \end{pmatrix} = \mu \begin{pmatrix} a \\ b \end{pmatrix}. \tag{5.98}$$

Solving the above nonlinear equations together with the complete particle conservation condition $|a|^2 + |b|^2 = 1$, we readily obtain the chemical potential μ and the eigenstate (a, b). The eigenenergies can be derived according to the relationship $\varepsilon = \mu - c(|a|^4 + |b|^4)/2$, and their dependence on the input parameters is plotted in Fig. 5.21 for the cases of linearity, weak nonlinearity, and strong nonlinearity. Striking phenomena are induced by the introduction of nonlinearity: First, the structure of the energy levels changes dramatically. In the linear case, two energy levels are symmetric about a horizontal axis [see Fig. 5.21a]. However, the symmetry breaks down in the presence of nonlinearity, and a new branch of eigenenergies emerges. In the case of weak nonlinearity [see Fig. 5.21b], two levels exist at the two ends and near the peak of the field pulse; three energy levels exist in the other regime. When the nonlinearity is strong [see Fig. 5.21c], apart from at the two ends, three energy levels exist. Second, the eigenstates of the middle level (e.g., those denoted by the dashed line in Fig. 5.21) are unstable. This finding is evaluated by investigating the Hamiltonian-Jacobi matrix obtained by linearizing (5.98) around the eigenstates. The eigenvalues of the Hamiltonian-Jacobi matrix can be real, complex or purely imaginary. Only purely imaginary eigenvalues correspond to stable states; others indicate unstable ones [49].

The changes of the energy level in the presence of nonlinearity are expected to strongly affect the quantum transition between levels. However, because of the degeneracy of our states of interest of (1, 0) and (0, 1), from the above diagram of the energy levels, we cannot draw a definite conclusion about the adiabatic evolution of the initial state (1, 0). In the following, we introduce an equivalent classical

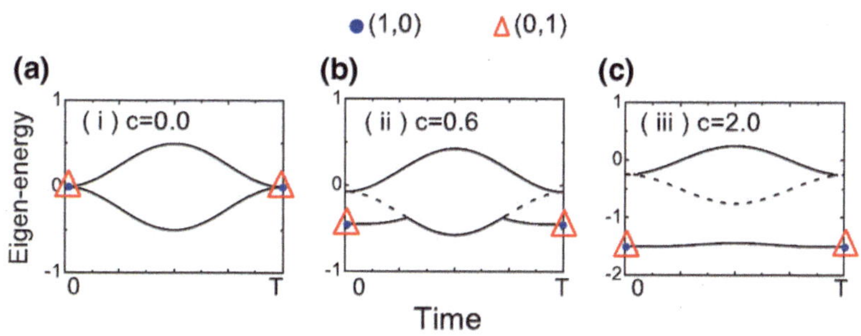

Fig. 5.21 Typical eigenenergy level structure in the **a** linear case, **b** weak nonlinearity case, and **c** strong nonlinearity case. The dashed lines correspond to the unstable eigenstates

Hamiltonian instead, and by analyzing its phase space, we achieve insight into the adiabatic evolution of the above nonlinear quantum system.

Following the theoretical methodology proposed in [4], we construct the effective classical Hamiltonian by introducing two quantities: the probability of particles staying in the (0, 1) mode, $p = |b(t)|^2$, and the relative phase of the two modes, $\theta = \theta_a - \theta_b$. These quantities form a pair of canonical variables of the following classical Hamiltonian and satisfy the canonical equations, i.e., $d\theta/dt = \partial H/\partial s$, $ds/dt = -\partial H/\partial\theta$, and

$$H = v\sqrt{p(1-p)}\cos\theta - \frac{c}{4}(2p-1)^2. \tag{5.99}$$

The above classical system is capable of completely describing the dynamic properties of the nonlinear quantum Rosen-Zener system (5.93) on a projective Hilbert space [4]. Its fixed points, i.e., energy extrema of the classical Hamiltonian, correspond to the quantum eigenstates. For example, in Fig. 5.22, the stable elliptic fixed point P_1 corresponds to the upper level of Fig. 5.21. The energies associated with P_2 and P_4 are identical; therefore, they correspond to the same energy level, i.e., the lower one in Fig. 5.21. The saddle point P_3 is unstable, corresponding to the middle level denoted by a dashed curve in Fig. 5.21. The adiabatic evolution of the quantum eigenstates can therefore be evaluated by tracing the shift of the classical fixed points in the phase space when the parameter v varies slowly in time [3].

The analytic expressions of the fixed points are obtained from $dp/dt = 0$ and $d\theta/dt = 0$:

$$\theta^* = 0, \pi, \tag{5.100}$$

$$\frac{v(1-2p^*)}{2\sqrt{p^*(1-p^*)}}\cos\theta^* + c(1-2p^*) = 0. \tag{5.101}$$

The number of fixed points depends on the instantaneous coupling strength v and the nonlinear parameter c. If $c/v < 1$, there exist only two fixed points $(\theta^*, p^*) = (0, 1/2), (\pi, 1/2)$. However, when $c/v > 1$, there exist four fixed points [P_1, P_2, P_3, and P_4 in Fig. 5.22a and b]: $(\theta^*, p^*) = (0, 1/2), (\pi, 1/2), [\pi, \frac{1}{2}(1 \pm \sqrt{1 - v^2/c^2})]$. One of them ($P_3$) is a saddle point, while the other three (P_1, P_2, and P_4) are all elliptic points corresponding to the local maximum (P_1) and minima (P_2 and P_4) of the classical Hamiltonian.

When we increase v, P_2, P_3, and P_4 merge into a new stable fixed point P_{234} in a regime satisfying the condition $v/c > 1$, as shown in Fig. 5.22a → b → c → d. An interesting question arises when the parameter v decreases to less than c again: Which point does the state follow when P_{234} bifurcates into P_2, P_3, and P_4 (Fig. 5.22d → c → b → a)? The state that follows only P_2 corresponds to zero adiabatic transition probability, whereas the state that follows only P_4 corresponds to a complete population transfer. This classical picture could explain why we see a rectangular pattern in Fig. 5.20.

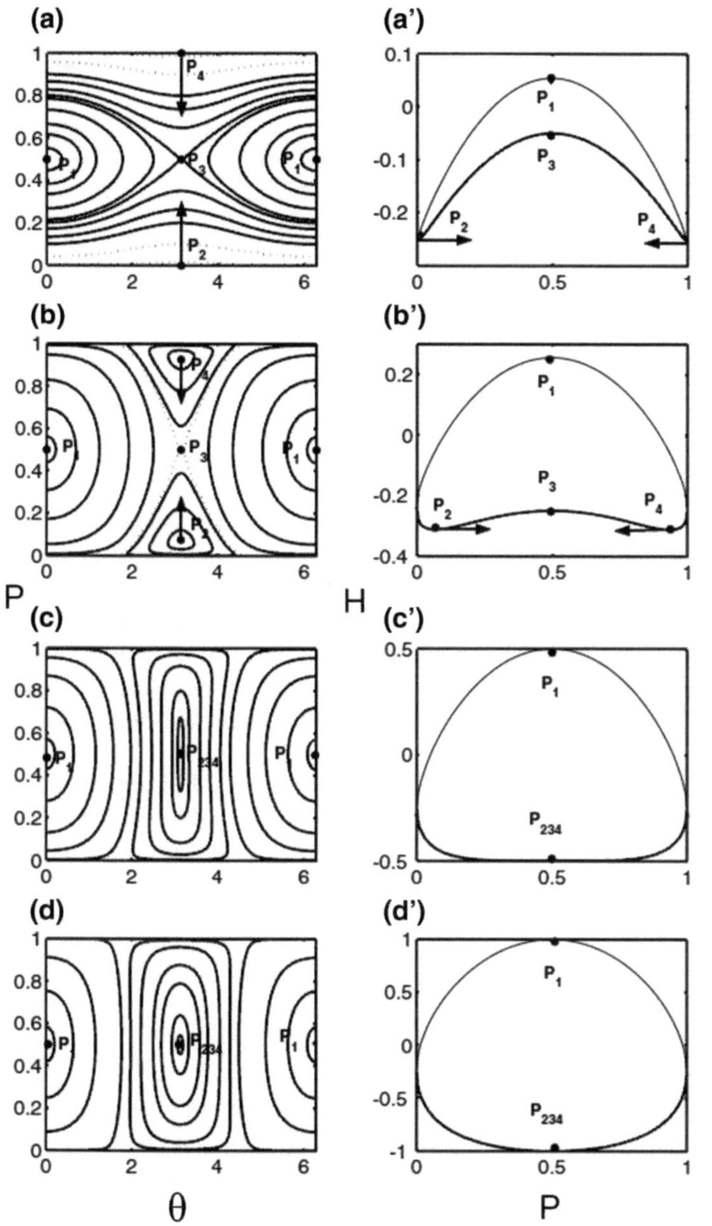

Fig. 5.22 Evolution of the phase space motions as c/v changes adiabatically: **a** 10, **b** 2, **c** 1, **d** 0.5. The second column is the corresponding energy curve at $\theta = 0$ (thin line) and π (heavy line). The arrows indicate the shifting direction of the fixed points as v increases

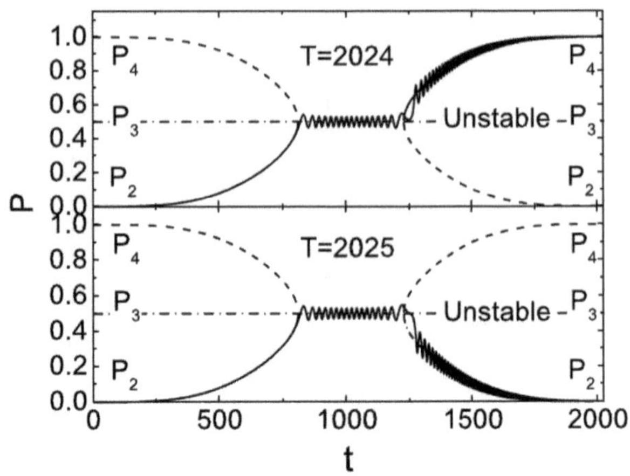

Fig. 5.23 Evolution of the particles and the fixed points at $c/v_0 = 0.9$. Entirely different transition probabilities are observed for a slight variation of the scanning period: **a** $T = 2024$ and **b** $T = 2025$

The above scenario is further supported by directly tracing the evolution of particles, as shown in Fig. 5.23, where we also demonstrate the temporal evolution of the fixed points. In the early stage, the state faithfully follows the fixed point P_2. The state starts to show a small oscillation when the fixed points P_2, P_3, and P_4 merge. Subsequently, the state follows either the fixed point P_2 or P_4. Notably, at a certain parameter value, a slight change in the period T can thoroughly change the final transition probability, a signature of the appearance of a bistable state [see Figs. 5.22c' → b'].

Our results show that whether the state follows P_2 or P_4 is determined by the total dynamical phase accumulated during the oscillation motion around P_{234}, i.e., from t^* to t^{**}. Here, t^* is the time when P_2, P_3, and P_4 merge into P_{234}, and t^{**} is the moment when P_{234} bifurcates into P_2, P_3 and P_4. The results are obtained by setting $c = v(t^*) = v(t^{**})$,

$$\begin{aligned} t^* &= \frac{T}{\pi}\sin^{-1}\sqrt{\frac{c}{v_0}}, \\ t^{**} &= T - t^*. \end{aligned} \tag{5.102}$$

To obtain the total phase, we first need to calculate the instantaneous frequency that characterizes the oscillations around the fixed point. To this end, we expand the classical Hamiltonian around the fixed point with $p = 1/2 + \delta_p$ and $\theta = \pi + \delta_\theta$, leading to

$$\delta H = \frac{1}{4}(v - c)\delta_p^2 + \frac{1}{4}v\delta_\theta^2, \tag{5.103}$$

ignoring the higher-order terms. The instantaneous frequency is then derived as

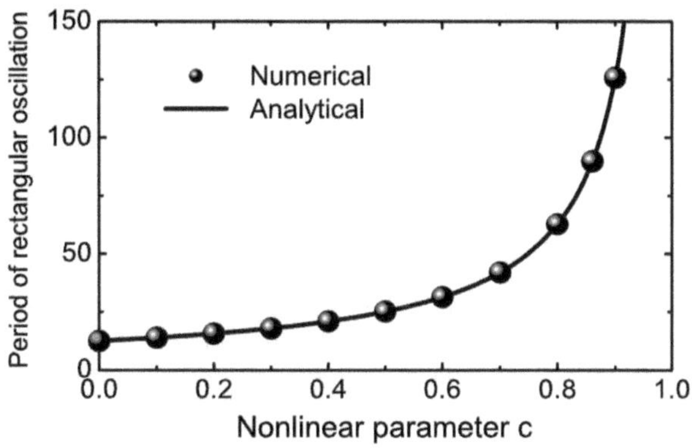

Fig. 5.24 Periods of the rectangular oscillation as shown in Fig. 5.20 under the adiabatic limit. The numerical results are obtained by directly observing Fig. 5.20, while the analytical curve is the plot of function $4\pi/(v_0 - c)$, where $v_0 = 1$. The results agree well with each other

$$\omega(t) = \frac{1}{2}\sqrt{v(t)[v(t) - c]}. \tag{5.104}$$

Integrating $\omega(t)$ from t^* to t^{**} gives the total phase

$$\varphi = \int_{t^*}^{t^{**}} \omega(t)dt = \frac{v_0 - c}{4}T. \tag{5.105}$$

This expression indicates that the total phase increases linearly with the scanning period. A π-value change in the phase changes the choice of the state to follow either fixed point P_2 or P_4. Thus, the period of the rectangular oscillation observed in Fig. 5.20 under the adiabatic limit can be expressed as

$$T_{rectangular} = \frac{4\pi}{v_0 - c}. \tag{5.106}$$

To verify the above theory, we have numerically solved the nonlinear Schrödinger equation for a wide range of parameters. The comparison between the analytical results and the numerical data in Fig. 5.24 shows good agreement.

The complete suppression of a quantum transition under the adiabatic limit in Fig. 5.20d can also be explained from the above picture. We briefly state this explanation as follows. For the strong nonlinearity associated with $c/v_0 > 1$, the phase space evolution only follows Fig. 5.22a $\rightarrow$ b $\rightarrow$ a as v increases and decreases. During the process, no collision between the fixed points occurs. Thus, the state initially populated on P_2 can safely remain on the fixed point and ultimately return smoothly to the original state. Thus, no transition is observed [48].

5.4 Nonlinear Ramsey Interferometry

The technique of Ramsey interferometry with separated oscillating fields was first proposed to investigate molecular beam resonance [50]. The key feature of the observed Ramsey pattern in the frequency domain is that the width of the central peak is determined by the inverse of the time taken by the particle to cross the intermediate drift region [51]. Indeed, the Ramsey interference experiments can be operated either in the time domain with temporally separated pulses and a fixed particle or in the space domain with spatially separated fields and a moving particle [52]. The Ramsey interferometric method provides the basis of atomic fountain clocks that now serve as time standards [53, 54] and has stimulated rapid advancement in the field of precision measurements in atomic physics. Since applying the laser cooling techniques to trapped atoms, the atomic interferometers with cold atoms have been used to measure rotation [55], gravitational acceleration [56, 57], the atomic fine-structure constant [58], the atomic recoil frequency [59], and atomic scattering properties [60], to name only a few applications.

On the other hand, the experimental realization of the Bose-Einstein condensate (BEC) in a dilute atomic gas [27, 30, 61] brings a fascinating opportunity for the purpose of precision measurement due to the very slow atoms and changes the prospects of frequency standards entirely. Recently, Ramsey fringes between atoms and molecules in the time domain were observed by using a trapped BEC of ^{85}Rb atoms [62] in an experiment. This approach offers the possibility of the precise measurement of the binding energy of the molecular state in BECs [63, 64]. With the development of atomic interferometry techniques, researchers are seeking to exploit new interferometric methods using trapped BECs [65, 66]. With the emergence of nonlinear interaction between the coherent ultracold atoms, BECs show intriguing nonlinear tunneling and interference properties that are distinct from those of the traditional quantum systems. Motivated by the discussion on the nonlinear Rosen-Zener (RZ) transition [48], we conceptually construct in this section a nonlinear Ramsey interferometer by applying a sequence of two identical nonlinear RZ tunneling processes (i.e., RZ pulses). In our interferometry scheme, two RZ pulses are separated by an intermediate holding period of variable duration; by varying the holding period, we have observed diversiform Ramsey interference patterns contrasting with the standard Ramsey fringes. Using a simple nonlinear two-mode model, we thoroughly discuss the physics underlying the interference patterns both numerically and analytically. We show that the frequency of the nonlinear Ramsey fringes exactly reflects the strength of nonlinearity in addition to the asymmetry of the system. This observation suggests a potential application in calibrating atomic parameters, such as the scattering length and energy spectrum, via measuring the frequency of Ramsey fringes [67].

We consider that a condensate, for example, ^{87}Rb atoms in a magnetic trap, is driven by microwave coupling into a linear superposition of two different hyperfine states. Since the microwave source injects very large numbers of photons, the electromagnetic field can be treated as a completely classical object. We can then identify the two basis states $|1\rangle$ and $|2\rangle$ of the $(F = 1, m_F = -1)$ and $(F = 2, m_F = +1)$ hyperfine states, respectively. We denote the corresponding boson creation (annihilation) operators $\hat{a}^\dagger$ $(\hat{a})$ and $\hat{b}^\dagger$ $(\hat{b})$. Considering that the laser fields are time dependent, we treat the Hamiltonian in the rotating frame, that is, the frame in which the laser field is constant over the time of the pulse. Under this condition, the operators $\hat{J}_x = (\hat{a}^\dagger\hat{b} + \hat{b}^\dagger\hat{a})/2$, $\hat{J}_y = i(\hat{a}^\dagger\hat{b} - \hat{b}^\dagger\hat{a})/2$, and $\hat{J}_z = (\hat{a}^\dagger\hat{a} - \hat{b}^\dagger\hat{b})/2$ can form a complete set of number-conserving Hermitian operators for the system, and the Hamiltonian can be expressed as a function of these operators. Within the standard rotating-wave approximation, the Hamiltonian describing the transition between the two internal states can be written [68] as

$$\hat{H} = -\gamma\hat{J}_z - c\hat{J}_z^2 + v\hat{J}_x, \tag{5.107}$$

where $\gamma = -\delta + (4N\pi\hbar^2/m)(a_{11} - a_{22})\eta$ is the energy difference between two states characterizing the asymmetry of the system, $c = (2\pi\hbar^2/m)(a_{11} + a_{22} - 2a_{12})\eta$ is the nonlinear strength describing atomic interactions, and v denotes the coupling strength, which is proportional to the intensity of the near-resonant laser field. δ is the detuning of lasers from resonance; a_{ij} is the s-wave scattering amplitude of hyperfine species i and j; η is a constant of order 1 independent of the hyperfine index, relating to an integral of the equilibrium condensate wave function; N is the atomic number; and m is the mass of the atom.

To obtain the model (5.107), the single mode approximation (SMA) is applied, that is, the density profiles of two component condensates are assumed to be identical and to remain constant during temporal evolution. The validity of the SMA for the spinor-1 atomic condensate, such as ^{87}Rb, has been previously investigated [69]. The results showed that the SMA works well under typical experimental conditions for ^{87}Rb atoms.

In the limit of large particle number, the operators in the above field equations can be replaced by complex numbers; we thus obtain the following mean-field equations that effectively describe the evolution of the above two-component BEC system: $(\hbar = 1)$,

$$i\frac{d}{dt}\begin{pmatrix} a \\ b \end{pmatrix} = H(v)\begin{pmatrix} a \\ b \end{pmatrix}, \tag{5.108}$$

with the Hamiltonian

$$H(v) = \begin{pmatrix} \frac{\gamma}{2} + \frac{c}{2}(|b|^2 - |a|^2) & \frac{v}{2} \\ \frac{v}{2} & -\frac{\gamma}{2} - \frac{c}{2}(|b|^2 - |a|^2) \end{pmatrix}, \tag{5.109}$$

where a and b denote the amplitudes of probabilities for two components and the total probability $|a|^2 + |b|^2 = 1$. Using the above two-component BEC system, we are capable of realizing a nonlinear Ramsey interferometer in which the nonlinearity represents the interparticle interaction. The main structure of our nonlinear Ramsey interferometer is illustrated in Fig. 5.25, in which the variation of the coupling strength is governed by two Rosen-Zener pulses of the form

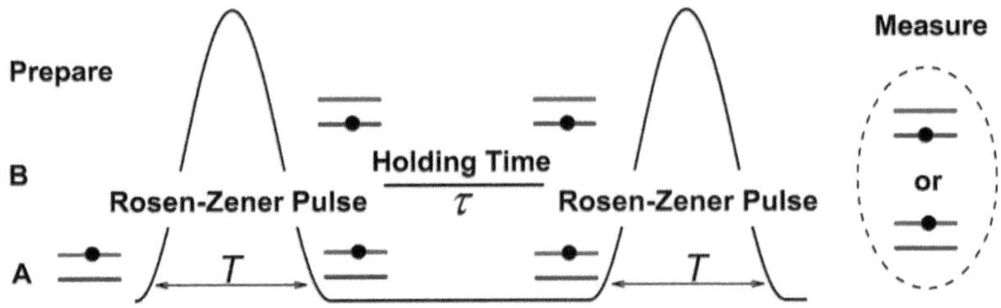

Fig. 5.25 (Color online) Schematic plot of a nonlinear Ramsey interferometer with a two-component trapped BEC in the time domain, starting with an RZ pulse, operating through a holding period, and ending with another RZ pulse

$$v(t) = \begin{cases} 0, & t < 0, \\ v_0 \sin^2\left(\frac{\pi t}{T}\right), & t \in [0, T], \\ 0, & t \in (T, T+\tau), \\ v_0 \sin^2\left(\frac{\pi t}{T}\right), & t \in [T+\tau, 2T+\tau], \\ 0, & t > 2T+\tau. \end{cases} \tag{5.110}$$

The above RZ pulses are characterized by the following parameters: v_0 is the maximum strength of the coupling, T is the scanning period of the RZ pulse, and τ is an alterable time interval between two pulses.

This scheme is analogous to a normal Ramsey interferometer in which the Ramsey pulses at the beginning and end of the sequence that couple the two components and redistribute the populations on each component are replaced by the so-called nonlinear RZ tunneling process [48]. The two tunneling processes are separated by a holding period, during which there is no coupling between the two components, and the BEC on each component evolves independently and only acquires different additional phases. In the course of the simulation experiments, the system is prepared in one internal state initially, and the final populations of atoms in each state is recorded when the second pulse is turned off. The measurements are repeated with a variable time interval τ. The final populations are sensitive to the phase difference that accumulates between the two components during the intermediate period; as a result, the Ramsey fringe pattern is expected to emerge in the time domain.

The nonlinear Schrödinger equations (5.108) that govern the temporal evolution of the two-component BEC system are solved numerically using a standard 4–5th-order Runge-Kutta algorithm. We set the initial condition $(a, b) = (1, 0)$ and take the maximum coupling strength as the energy scale, namely, $v_0 = 1$. The Ramsey fringe patterns are obtained by recording the final transition probability $|b|^2$ versus the holding time τ.

We begin our numerical simulations with $c = 0.6$ and $T = 1500$ for both the symmetric case $\gamma = 0$ and the asymmetric case $\gamma = 0.1$. With the emergence of nonlinearity, the Ramsey fringe pattern distinctly deviates from that of the linear case due to the dramatic changes of the transition dynamics. In this case, the system described in (5.108) is no longer analytically solvable. Our numerical results are displayed in Fig. 5.26, which shows that both nonlinearity and symmetry can affect the

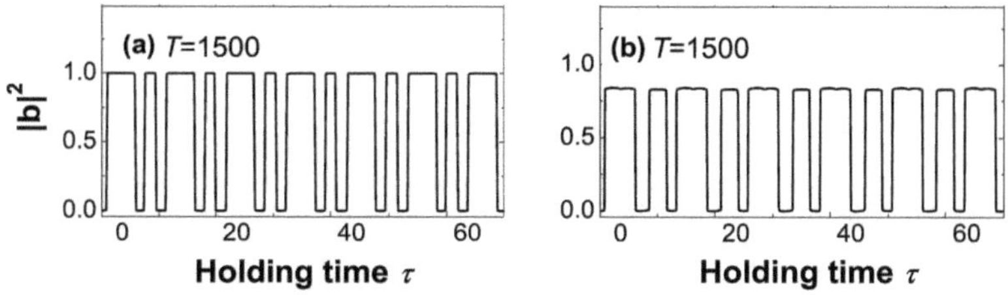

Fig. 5.26 (Color online) Ramsey fringe patterns for the **a** symmetric case and **b** asymmetric case with $c = 0.6$ and $T = 1500$

pattern and frequency of Ramsey fringes significantly. In our slow scanning case of $T = 1500$, the Ramsey fringe pattern exhibits a rectangular oscillation. This diversiform interference pattern is distinct from the normal Ramsey fringes of sinusoidal or cosinusoidal forms and is clearly generated by the nonlinear atomic interaction.

Here, we present a thorough analysis of these striking interference patterns. In practical experiments, in contrast to the oscillating amplitudes and shapes of the fringe patterns, the frequencies of the patterns are of more interest and can be recorded with relatively high resolution and contrast; therefore, we focus our theoretical analysis on the frequency property extracted from the Ramsey interference patterns through the Fourier transformation (FT). We find that the frequencies of patterns that are dramatically modulated by the interplay of nonlinearity and symmetry contain much information about the intrinsic properties of the BEC system.

Through investigating the nonlinear Ramsey patterns presented above, we find that the time scale of the period of the RZ pulse plays an important role in forming the striking patterns. The following discussion focuses only on the adiabatic limit, which refers to the case in which the RZ pulse is much slower than the intrinsic motion of the system, i.e., $T \gg 2\pi/v_0$.

For convenience, we introduce a phase shift $\phi(\tau)$ to describe the different phase accumulations between two components during the holding period. Considering that the two components evolve independently during this period, we obtain $\phi(\tau) = |\gamma + cs|\tau$ from (5.108), where $s = |b(T)|^2 - |a(T)|^2$ denotes the population difference between two components when the first pulse has been turned off. This phase shift is proportional to the holding time. The angular frequency of the Ramsey fringes is expected to be

$$\omega = |\gamma + cs|. \tag{5.111}$$

This result implies that the frequency of Ramsey fringes is entirely determined by the population difference s and the parameters γ and c.

One important phenomenon in the adiabatic limit case is that the FT of the Ramsey fringes reveals multiple frequency components, namely, $\omega = n\omega_0$, where ω_0 is the fundamental frequency (i.e., basic or first frequency) of the fringes and n is a positive integer. We interpret this result in terms of the interplay between nonlinearity ascribed to the interatomic interaction and the coupling energy from the external laser

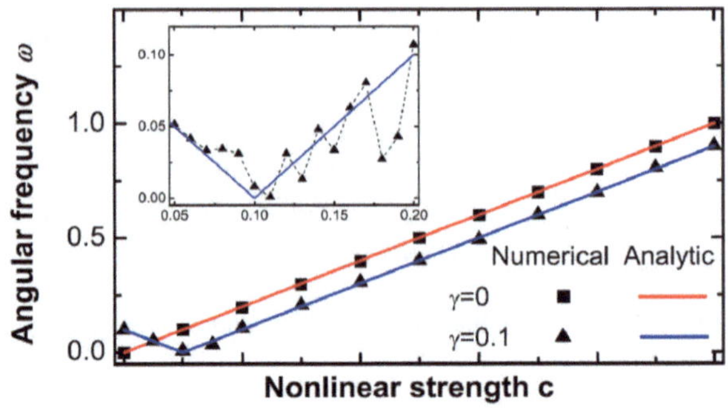

Fig. 5.27 (Color online) Fundamental angular frequency of Ramsey fringes as a function of the nonlinear strength c for the adiabatic limit case. The inset demonstrates the details for $\gamma = 0.1$ with c from 0.05 to 0.2

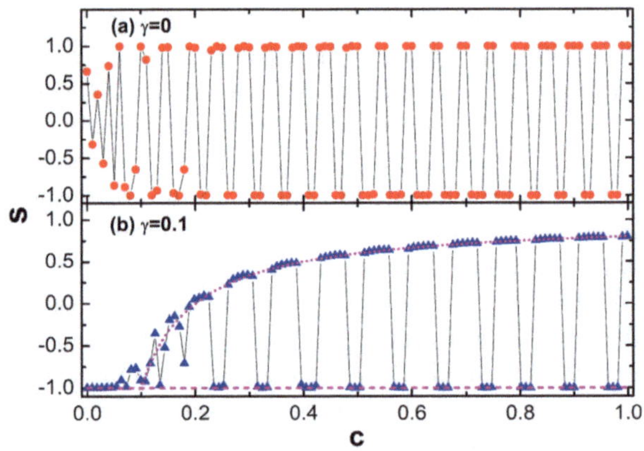

Fig. 5.28 (Color online) Population difference s versus nonlinear parameters from 0 to 1 for the symmetric case (red circles) and the asymmetric case (blue triangles) with $T = 1500$. The dotted and dashed lines refer to theoretical predictions from (5.116)

field. Figure 5.27 illustrates only the fundamental frequencies of Ramsey fringes for different nonlinear parameters.

An irregular fluctuation appears near the $c = \gamma$ region (see the inset in Fig. 5.27). We hypothesize that the adiabatic assumption is violated in this region. To test this hypothesis, we trace the population difference s after the first RZ pulse with the nonlinear parameter increasing. The results are presented in Fig. 5.28, which shows an irregular oscillation of s occurring in the region where $|\gamma - c|$ is also very small. With the nonlinear parameter increasing from 0.25 to 1, s jumps between the two points $+1$ and -1 in the symmetric case. However, for the asymmetric system, when $c > 0.35$, the value of s jumps between -1 and another unknown point. This is an intriguing quantum phenomenon, and the underlying physical reasons require further detailed discussion.

To explain the above peculiar phenomena, under the mean-field approximation, following [4], we introduce the relative phase $\theta = \theta_b - \theta_a$ and the population difference $s = |b|^2 - |a|^2$ as two canonical conjugate variables, enabling us to obtain an effective classical Hamiltonian

$$\mathscr{H} = -\left(\gamma + \frac{c}{2}s\right)s + v\sqrt{1 - s^2}\cos\theta. \tag{5.112}$$

This classical Hamiltonian can describe completely the dynamical properties of the system described in (5.108) [4]. The adiabatic evolution of the quantum eigenstates can be evaluated by tracing the shift of the classical fixed points in the phase space when the parameter v varies slowly in time [3]. According to [48, 70], for the symmetric system, we obtain the classical fixed points on the line $\theta^* = \pi$,

$$s^* = \begin{cases} 0, & c/v < 1, \\ 0, \pm\sqrt{1-(v/c)^2}, & c/v > 1. \end{cases} \tag{5.113}$$

We show the evolution of the fixed point $s^* = -1$ (P_2) in Fig. 5.29. The three fixed points in (5.113) are characterized by P_3, P_4, and P_2. The one saddle point P_3 ($s^* = 0$) and two elliptic points P_2 and P_4 correspond to one unstable state and two stable states, respectively. For $c = 0.8$, consistency between the dynamical evolution and adiabatic trajectory of P_2 is shown for both $T = 1500$ and $T = 20,000$. However, for $c = 0.03$, the evolution of fixed point P_2 shows a clear deviation from the adiabatic trajectory given by (5.113) at $T = 1500$ [see Fig. 5.29a], while the fixed point follows the adiabatic evolution at $T = 20,000$ [see Fig. 5.29b]. These phenomena indicate that the adiabatic condition cannot be satisfied for $c = 0.03$, where irregular fluctuation occurs at $T = 1500$ in Fig. 5.28. Therefore, we describe the adiabatic condition as follows:

$$T \gg \text{Max}\left[\frac{2\pi}{|\gamma - c|}, \frac{2\pi}{v_0}\right]. \tag{5.114}$$

Under this condition, as long as $\gamma \neq c$, the system evolves adiabatically if the scanning period is long enough even for small nonlinear parameters [70]. This condition can successfully explain the fluctuation in Figs. 5.27 and 5.28. Accordingly, we trace the fixed point P_2 in the asymmetric case (see Fig. 5.30) using the same parameter T as in Fig. 5.29. We observe the similar feature that satisfactory adiabatic evolution for $c = 0.8$ and nonadiabatic evolution for $c = 0.18$ occur in close vicinity to the zero-energy resonance ($\gamma = c$) with $T = 1500$. In the asymmetric case, another interesting phenomenon is that the destination of the evolution of the fixed point is not definite, i.e., two different target states exist [see Figs. 5.28b and 5.30b]. We interpret this situation through a comprehensive physical analysis below.

For the adiabatic limit case, the energy of the system for both the symmetric and asymmetric cases is no longer conserved during the entire evolution process; however, at the beginning and end of the evolution, the corresponding energies of the system keep the same value,

$$\mathscr{H}(s = -1, t = 0) = \mathscr{H}(s^*, t = T). \tag{5.115}$$

In our scheme, for both $t = 0$ and $t = T$, the coupling parameter $v = 0$. Thus, we can obtain the final state of the system from (5.112) and (5.115):

$$s^* = \begin{cases} -1, & \gamma > c, \\ -1, 1 - 2\gamma/c, & 0 < \gamma < c. \end{cases} \tag{5.116}$$

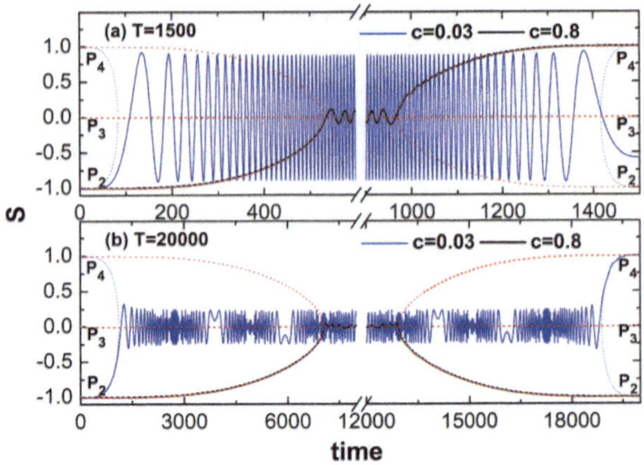

Fig. 5.29 (Color online) Comparison between the dynamical evolution (solid line) and the adiabatic evolution (dotted line) of fixed points for the symmetric case with different T: **a** 1500 and **b** 20, 000. The blue line and black line refer to $c = 0.03$ and $c = 0.8$, respectively. The blue dotted line and red dotted line show the corresponding adiabatic evolution obtained from (5.113)

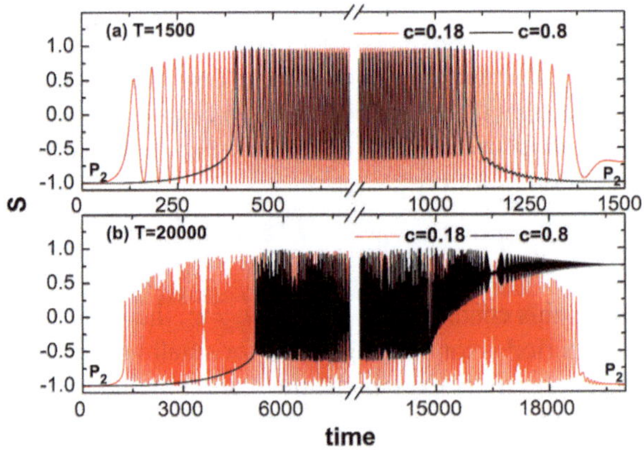

Fig. 5.30 (Color online) Evolution of fixed points for the asymmetric case under different T: **a** 1500 and **b** 20, 000

This result implies that at the end of the adiabatic evolution, the system has two states to choose from when $c > \gamma$ for this case. One choice is back to the initial state $s^* = -1$, and the other choice is located on another state with energy identical to that of the initial state $s^* = 1 - 2\gamma/c$. However, the latter choice restricts the population to $|b|^2 = \gamma/c$. In other words, quantum tunneling for the asymmetric case requires the atom number on another state to not exceed $N\gamma/c$ (where N is the total number of atoms). We use the above analysis to evaluate our numerical results in Fig. 5.28b and find satisfactory consistency. According to this analytic prediction, in the adiabatic limit case, the final value of s is -0.11 or -1 for $c = 0.18$ and 0.75 or -1 for $c = 0.8$ in Fig. 5.30; these results strongly support our numerical results.

Following the above analysis, we can obtain the analytic expression of the fundamental frequency of Ramsey fringes in the adiabatic limit,

$$\omega = |\gamma - c|. \tag{5.117}$$

The results, which show a perfect linear relation for both the symmetric and asymmetric cases, are consistent with our numerical results (see Fig. 5.27).

Our nonlinear Ramsey interferometer scheme can also be realized using the BECs with a double-well potential. This BEC system, under the mean-field approximation, is described by the Gross-Pitaevskii equation (GPE):

$$i\hbar\frac{\partial\Psi(r,t)}{\partial t}=\left(-\frac{\hbar^2}{2m}\nabla^2+V(r)+U_0|\Psi(r,t)|^2\right)\Psi(r,t), \tag{5.118}$$

where $U_0=4\pi\hbar^2 a_s N/m$, with m being the atomic mass and a_s being the s-wave scattering length of the atoms. The wave function can be described by a superposition of two states that localize in each well separately as [28] $\Psi(r,t)=\psi_1(t)\phi_1(r)+\psi_2(t)\phi_2(r)$. The spatial wave function $\phi_i(r)$ $(i=1,2)$ which describes the condensate in each well, can be expressed in terms of symmetric and antisymmetric stationary eigenstates of the GPE, and these two wave functions satisfy the orthogonality condition $\int\phi_1(r)\phi_2(r)dr=0$ and normalized condition $\int|\phi_i|^2dr=1$. Consider the weakly linked BEC; the dynamic behavior of the system can be described by the Schrödinger equation with the following Hamiltonian:

$$H=\begin{pmatrix}\varepsilon_1^0+c_1|\psi_1|^2 & K\\ K & \varepsilon_2^0+c_2|\psi_2|^2\end{pmatrix}, \tag{5.119}$$

where $\varepsilon_i^0=\int[(\hbar^2/(2m))|\nabla\phi_i|^2+|\phi_i|^2V(r)]dr$ $(i=1,2)$ is the zero-point energy in each well, $\Delta\varepsilon=\varepsilon_1^0-\varepsilon_2^0$ is the energy bias, $c_i=U_0\int|\phi_i|^4dr$ denotes the atomic self-interaction, and $K=\int[(\hbar^2/(2m))(\nabla\phi_1\nabla\phi_2)+\phi_1V(r)\phi_2]dr$ stands for the amplitude of the coupling between two wells.

For example, consider the one-dimensional case; we can express the potential of our system as $V(x)=m\omega^2x^2/2+v_0\mathrm{e}^{-x^2/(2d)}+fx$, where d is the double-well separation in the x direction. This optical double-well potential can be created by superimposing a blue-detuned laser beam upon the center of the magnetic trap [27], and the difference of the zero-point energy between two wells or trap asymmetry characterized by f can be found by a magnetic field, a gravity field or light shifts [71]. The atomic interaction c can be adjusted flexibly by Feshbach resonance, and the barrier height K can be effectively controlled by adjusting the intensity of the blue-detuned laser beam.

5.5 Nonlinear Atom-Molecule Conversion

5.5.1 Bosonic Atoms to Bosonic Molecules

The production of ultracold diatomic molecules in bosonic systems is an exciting area of research with important applications ranging from the production of molecular Bose-Einstein condensates (BECs) [72] to the study of chemical reaction dynamics

[73]. A widely used production technique involves the association of ultracold atoms into very weakly bound diatomic molecules by applying a time-varying magnetic field in the vicinity of a Feshbach resonance [74, 75]. The underlying conversion dynamics are typically described by the Landau-Zener (LZ) model [22]. In this model, the Feshbach molecule production is discussed under a two-body configuration in which a single pair of atoms is converted to a molecule at an avoided crossing between the atomic energy level and the molecular energy level as the molecular energy is lifted by an applied linearly sweeping magnetic field. Thus, the molecule production efficiency is expected to be of the exponential Landau-Zener type [76, 77].

In the above two-body model [76, 77] and its various many-body extensions [78–80], however, the interactions between particles (such as atom-atom, atom-molecule, and molecule-molecule interactions) are completely ignored. In this section, we consider a generalized many-body model for Feshbach molecule formation that includes atom-atom, atom-molecule, and molecule-molecule interactions. We show that the picture of two-body molecule production depicted by the Landau-Zener model is significantly altered by the particle interactions. In the adiabatic limit, we derive a formula for the upper bound of conversion efficiency when the interaction strength is larger than a critical value. Our theory predicts a significant role of the particle interactions in atom-molecule conversion when the atomic density is high and the Feshbach resonance width is narrow [81].

The two-channel bosonic model Hamiltonian that includes the atom-atom, atom-molecule, and molecule-molecule interactions takes the following form [82]:

$$\begin{aligned}\hat{H} = & \frac{u_{aa}}{V}\hat{a}^\dagger\hat{a}^\dagger\hat{a}\hat{a} + \frac{u_{bb}}{V}\hat{b}^\dagger\hat{b}^\dagger\hat{b}\hat{b} + \frac{u_{ab}}{V}\hat{a}^\dagger\hat{a}\hat{b}^\dagger\hat{b} \\ & + \varepsilon_a\hat{a}^\dagger\hat{a} + \varepsilon_b\hat{b}^\dagger\hat{b} + \frac{\omega}{\sqrt{V}}(\hat{a}^\dagger\hat{a}^\dagger\hat{b} + \hat{b}^\dagger\hat{a}\hat{a}). \end{aligned} \tag{5.120}$$

This model was proposed by Santos et al. [82] to investigate the Josephson oscillation and self-trapping phenomena of atom-molecule conversion systems. In (5.120), $\hat{a}^\dagger$ ($\hat{b}^\dagger$) and ε_a (ε_b) are the creation operator and chemical potential for the atomic (molecular) mode, respectively. In experiments, the magnetic field is linearly swept, $B(t) = (dB/dt)t$, and crosses the Feshbach resonance at B_0; thus, $2\varepsilon_a - \varepsilon_b = \mu_{co}[B(t) - B_0]$. Here, μ_{co} is the difference between the magnetic moments of a molecule and a pair of separated atoms. $\omega = \sqrt{4\pi\hbar^2 a_{bg}\Delta B\mu_{co}/m}$ denotes the amplitude for the interconversion of atoms and molecules due to the Feshbach resonance, in which m is the mass of a bosonic atom, a_{bg} is the background scattering length, and ΔB is the width of the resonance. $u_i = 2\pi\hbar^2 a_i/m_i$ ($i = aa, bb, ab$) are the atom-atom, molecule-molecule, and atom-molecule particle interactions. Here, a_i and m_i denote the background scattering length and the reduced mass, respectively, i.e., $a_{aa} = a_{bg}$, $a_{ab} \simeq 1.2a_{bg}$, $a_{bb} \simeq 0.6a_{bg}$ [83], $m_{aa} = m/2$, $m_{ab} = 2m/3$, and $m_{bb} = m$. We introduce N and V to denote the initial atom number and the quantized volume, respectively; therefore, $n = N/V$ is the mean density of the initial bosonic atoms.

In the mean-field limit, i.e., $N \to \infty$, the quantum fluctuation is negligible. One can replace all the quantum operators with c numbers; thus, the Heisenberg equations for operators $\hat{a}$ and $\hat{b}$ are cast as the following nonlinear Schrödinger equation:

$$i\frac{d}{dt}\begin{pmatrix} a \\ b \end{pmatrix} = H\begin{pmatrix} a \\ b \end{pmatrix}, \tag{5.121}$$

where

$$H = \begin{pmatrix} 2U(2|b|^2 - |a|^2) + \Delta & 4\Omega a^* \\ 2\Omega a & -4U(2|b|^2 - |a|^2) - 2\Delta \end{pmatrix}, \tag{5.122}$$

with $U = n(u_{ab}/2 - u_{aa} - u_{bb}/4)/4$, $\Delta = (2\varepsilon_a - \varepsilon_b + 2nu_{aa} - nu_{bb}/2)/4$, and $\Omega = \sqrt{n}\omega/2$. The total population is normalized to unity, i.e., $|a|^2 + 2|b|^2 = 1$.

We first show how the nonlinear interactions lead to the deformation of the eigenenergy levels. The eigenstates of the system satisfy

$$H\begin{pmatrix} a \\ b \end{pmatrix} = H\begin{pmatrix} \mu & 0 \\ 0 & 2\mu \end{pmatrix}\begin{pmatrix} a \\ b \end{pmatrix}. \tag{5.123}$$

Note that a diatomic molecule is composed of two atoms; thus, the factor 2 appears before the chemical potential for the molecular mode. Solving the above nonlinear equations together with $|a|^2 + 2|b|^2 = 1$, we readily obtain the chemical potential μ and the eigenstate (a, b). The eigenenergies can be derived according to the relationship $\varepsilon = \mu/2 + \mu|b|^2 + \Delta|a|^2/2 + 4U|b|^4 - 2U|a|^2|b|^2$. Their dependence on the parameters is plotted in Fig. 5.31. In the linear case [$U = 0$, Fig. 5.31a], the energy levels are symmetric such that they are invariant under a rotation about the zero point by 180°. Only two eigenstates exist when $|\Delta|$ is sufficiently large: one for the atomic mode and the other for the molecular mode. When $|\Delta|/\Omega < \sqrt{2}$, there is an additional eigenstate represented by the dotted line in Fig. 5.31. This eigenstate is dynamically unstable. With the appearance of nonlinear interaction, the symmetry of the energy levels breaks down. For the weak nonlinearity case $U/\Omega < \sqrt{2}/4$ [see Fig. 5.31b], the energy level structure is very similar to that of the linear case except for a slight shift. However, when $U/\Omega > \sqrt{2}/4$, a loop structure appears at the lower energy level. The loop expands as U increases, and the gap between the upper and lower energy levels becomes increasingly narrow. Such deformation of energy levels consequently leads to very different conversion dynamics.

Consider the adiabatic evolution of the system starting from the atomic mode at the left side of the lower energy level. When U is small, e.g., in Fig. 5.31a, the evolution of the system follows the solid line, converting all atoms into molecules. However, when $U/\Omega > \sqrt{2}/4$, as in Fig. 5.31c, the system moves steadily from the left side to the critical point C. Subsequently, no movement options remain except to jump to the upper and lower levels. As this fraction of atoms tunnels to the upper level, they are not converted into molecules. The situation becomes more extreme when U is very large: the critical point is much closer to the upper level and far from the lower one; thus, the system jumps to the upper level more easily, see Fig. 5.31d. As a result, almost all atoms cannot be converted into molecules.

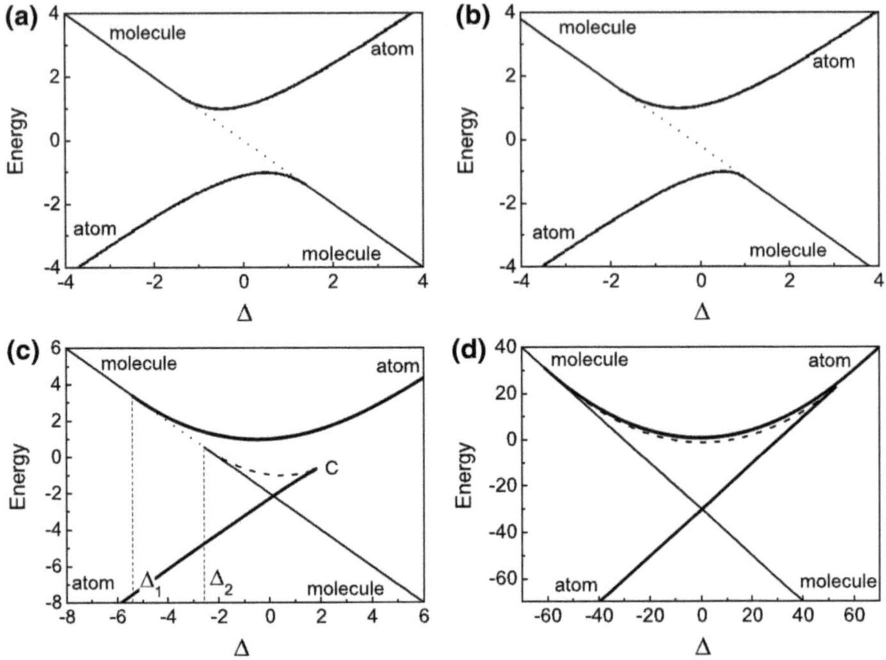

Fig. 5.31 Adiabatic energy levels for different nonlinear interaction strengths: **a** $U = 0$, **b** $U = 0.2$, **c** $U = 2$, and **d** $U = 30$. In all cases, $\Omega = 1$, the solid lines represent stable eigenstates, and the dotted lines between $\Delta_1 = U - \sqrt{2}$ and $\Delta_2 = U + \sqrt{2}$ correspond to unstable states. When $U > \sqrt{2}/4$, a loop structure appears at the lower energy level. The loop expands as U increases

The above simple analysis is confirmed by our numerical results, which are plotted in Fig. 5.32. In our calculations, the 4–5th-order Runge-Kutta step-adaptive algorithm is used to solve the nonlinear Schrödinger equation (5.121). The conversion efficiency as a function of the sweep rate $\alpha = d\Delta/dt$ is plotted and shows a monotonic decrease as the sweep rate increases. Even in the absence of particle interactions, i.e., $U/\Omega = 0$, the conversion efficiency calculated from our many-body model is quite different from that in the two-body Landau-Zener theory. For example, in the sudden limit of $\alpha/\Omega^2 \gg 1$, the former is almost twice the latter. In the presence of particle interactions, the atom-molecule conversion efficiency is further altered. When the dimensionless interaction parameter is positive, the atom-molecule conversion is suppressed relative to the case of $U/\Omega = 0$, whereas the conversion efficiency is enhanced when this dimensionless parameter is negative. From the explicit expression of the effective interaction parameter, we observe that the repulsive atomic interaction leads to a negative parameter U/Ω, while the attractive atomic interaction corresponds to a positive parameter. For large positive interaction parameters, even in the adiabatic limit of $\alpha/\Omega^2 \rightarrow 0$, the atoms cannot be totally converted into molecules. The saturation of conversion efficiency has already been discussed qualitatively in [84], without, however, including a quantitative analysis. Below, following

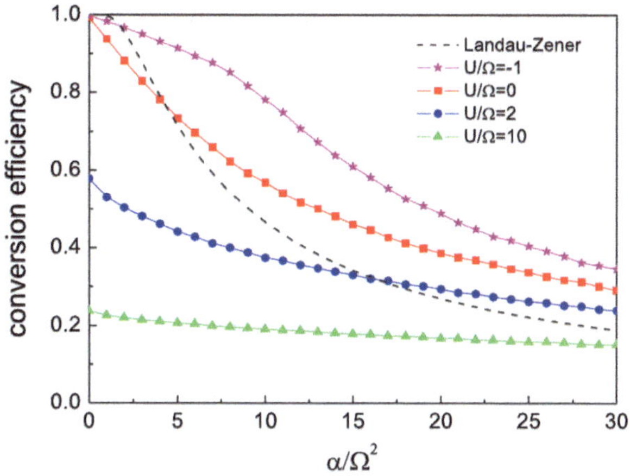

Fig. 5.32 Conversion efficiency as a function of the sweep rate α/Ω^2 for various interaction parameters. For comparison, we also plot the Landau-Zener-type formula $\chi = 1 - \Gamma_{lz} = 1 - e^{-2\pi\Omega^2/\alpha}$ [22, 79]; here, the sweep rate is defined as $\alpha = d\Delta/dt = \mu_{co}(dB/dt)/4$

the methodology of [3], we present selected analytical calculations in the adiabatic limit corresponding to $\alpha/\Omega^2 \ll 1$.

When introducing the canonical transformation $S = |a|^2 - 2|b|^2$ and $\theta = 2\theta_a - \theta_b$, where $\theta_a = \arg(a)$ is the phase of the atomic mode and $\theta_b = \arg(b)$ is the phase of the molecular mode, the quantum system is equivalent to the following classical Hamiltonian:

$$\mathscr{H} = -2US^2 + 2\Delta S + 2\Omega(1+S)\sqrt{1-S}\cos\theta. \qquad (5.124)$$

The canonical variables satisfy $dS/dt = -\partial H/\partial\theta$ and $d\theta/dt = \partial H/\partial S$. The above Hamiltonian is the same as that obtained in [82]. As previously noted [84], this Hamiltonian contains properties distinct from those of the "standard" nonlinear Landau-Zener Hamiltonian of [3]. First, the nonlinearity in Hamiltonian (5.122) arises not only from the particle interactions but also from the fact that two atoms are needed to form a molecule. In addition, the mean-field motion is restricted on a teardrop-shaped surface [84, 85] rather than the surface of a Bloch sphere. As we show later, the interplay of these features leads to very different conversion dynamics.

In the adiabatic limit where the external field varies slowly relative to the intrinsic motion of the system, the conversion dynamics is entirely determined by the phase-space structural evolution of the classical Hamiltonian (5.124); see, for example, [82, 84]. The fixed points (i.e., the energy extrema of the classical Hamiltonian) on the phase space correspond to the quantum eigenstates. According to the adiabatic theory [4, 33], when the energy bias Δ changes adiabatically, a closed orbit in the phase space remains closed, and the action $I = (2\pi)^{-1}\oint S d\theta$ stays invariant in time. The action equals the phase-space area enclosed by the closed orbit and is zero when the orbit shrinks to a fixed point.

For the case of $U/\Omega < \sqrt{2}/4$, the initial state is prepared at an elliptical point on the phase space. It evolves, following the elliptical point, from the boundary line of $S = 1$ to $S = -1$ as Δ increases. Consequently, all the atoms are converted into molecules, i.e., the conversion efficiency is $\chi = 1$.

However, for the case of $U/\Omega > \sqrt{2}/4$, the elliptical point collides with a saddle point when $\Delta = \Delta_c$. After this collision, the system enters a new orbit with $\mathscr{H} = \mathscr{H}_c$ and evolves adiabatically for $\Delta > \Delta_c$ according to the rule of constant action, which is now nonzero. This orbit ultimately evolves into a straight line of constant S. With these considerations, we can obtain the conversion efficiency in the adiabatic limit, namely, $\chi = 1 - I_c/2$.

To derive the explicit expression of χ, we first need to determine the critical point C. For this purpose, we note that the point C (with $\theta = \pi$) is a double root of $d\theta/dt = 0$; thus, $\partial(d\theta/dt)/\partial S|_{S_c} = -4U + \Omega(5 - 3S_c)/[2(1 - S_c)^{3/2}] = 0$. Once S_c is obtained, the critical energy bias Δ_c and orbit energy $\mathscr{H}_c$ can also be easily determined through their explicit expressions given in the foregoing paragraphs. The complete orbit passing through the critical point is given by $\cos\theta = f(S) = (\mathscr{H}_c + 2US^2 - 2\Delta_c S)/(2\Omega(1 + S)\sqrt{1 - S})$. Thus,

$$I_c = \frac{1}{\pi}\int_{S_{min}}^{S_c}\left[\pi - \arccos\left(\frac{\mathscr{H}_c + 2US^2 - 2\Delta_c S}{2\Omega(1 + S)\sqrt{1 - S}}\right)\right]dS + 1 + S_{min}. \quad (5.125)$$

Here, S_{min} can be determined by $f(S_{min}) = 1$. The above formula can be further simplified at the critical point of $U/\Omega \to \sqrt{2}/4$ and in the asymptotic regime of $U/\Omega \to \infty$ following [81]:

$$\chi = \begin{cases} 1 - 2.4\left(\frac{U}{\Omega} - \frac{\sqrt{2}}{4}\right)^2, & \frac{U}{\Omega} \to \frac{\sqrt{2}}{4}; \\ 1.2\left(\frac{U}{\Omega}\right)^{-2/3}, & \frac{U}{\Omega} \gg 1. \end{cases} \quad (5.126)$$

The above discussion is restricted to the attractive interaction case; nevertheless, it is straightforward to extend the discussion to the repulsive interaction case. In the latter case, the adiabatic energy levels are the 180° angular rotation of the levels presented in Fig. 5.31; accordingly, the conversion from atoms to molecules starting from the upper level at the right side in the repulsive case is equivalent to the atom-molecule conversion starting from the lower level at the left side in the attractive case when the sweeping direction of the magnetic field is reversed. Our calculation in Fig. 5.32 shows that for both the repulsive and attractive cases, the atom-molecule conversion efficiency is significantly affected by the particle interactions when $|U/\Omega|$ is similar to or larger than the critical value of $\sqrt{2}/4$.

5.5.2 *Fermionic Atoms to Bosonic Molecules*

Next, we discuss the Feshbach conversion of fermionic atom pairs to condensed bosonic molecules with a microscopic model that accounts for the repulsive interactions among all the particles involved [86]. The Feshbach resonance has become a focal point of research activities in the field of cold-atom physics [74, 75, 87, 88] after its first experimental observation in atomic gases [89]. Among these research activities, the production of diatomic molecules from Fermi atoms with Feshbach resonance is of special interest and has attracted a great deal of attention. First, this phenomenon is intrinsically an interesting one [90]; second, the phenomenon provides unique experimental access to Bardeen-Cooper-Schrieffer (BCS)–Bose-Einstein condensate (BEC) crossover physics [91–93]. To date, by slowly sweeping the magnetic field through the Feshbach resonance, samples of over 10^5 weakly bound molecules at temperatures of a few tens of nanokelvins have been produced from quantum degenerate Fermi gas [91, 94–96]. The Feshbach conversion is a complex process involving many fermionic atoms and bosonic molecules in a sweeping magnetic field that crosses a resonance. The theoretical description of the conversion efficiency as a function of the sweep rate, atomic mass, atomic density and temperature is still under development. The existing theories include the Landau-Zener (LZ) model of two-body molecular production [22, 33, 77] and its many-body extension at zero temperature [85, 97, 98], the phase-space density model [99], the equilibration model [100] and the quantum statistics model [101] at finite temperatures.

Here, we consider a microscopic model that describes the conversion from fermionic atoms to bosonic Feshbach molecules. We find that the two-body interactions can affect the Feshbach conversion efficiency: the repulsive interaction between molecules tends to enhance the conversion efficiency, whereas the other two repulsive interactions between atoms and between atoms and molecules suppress the efficiency. The role of the particle interactions is more significant for a narrow Feshbach resonance, where—in the adiabatic limit—the combined effect of these interactions can yield an upper limit of less than 100% on the conversion efficiency. This interaction-suppressed conversion efficiency is essentially the same as the broken adiabaticity induced by interaction in nonlinear LZ tunneling [2, 3].

To include all particle interactions, we extend the two-channel model in [101–104] and write the Hamiltonian as

$$
\begin{aligned}
H = & \sum_{k,\sigma} \varepsilon_k a^{\dagger}_{k,\sigma} a_{k,\sigma} + \left(\gamma + \frac{\varepsilon_b}{2}\right) b^{\dagger} b + \frac{U_b}{V_b} b^{\dagger} b^{\dagger} b b + \frac{U_a}{V_a} \sum a^{\dagger}_{k,\uparrow} a^{\dagger}_{-k,\downarrow} a_{k',\uparrow} a_{-k',\downarrow} \\
& + \frac{U_{ab}}{V_a} \sum_{k,\sigma} a^{\dagger}_{k,\sigma} a_{k,\sigma} b^{\dagger} b + \frac{g V_b}{V_a^{3/2}} \sum_{k} \left(b^{\dagger} a_{k,\uparrow} a_{-k,\downarrow} + a^{\dagger}_{-k,\downarrow} a^{\dagger}_{k,\uparrow} b\right), \qquad (5.127)
\end{aligned}
$$

where $\varepsilon_k = \hbar^2 k^2/2m_a$ is the kinetic energy of the atom, $\sigma = \uparrow, \downarrow$ denotes the two hyperfine states of the atom, $\varepsilon_b/2$ is the molecular energy, and $U_b = 4\pi\hbar^2 a_{bb}/m_b$ is the interaction between molecules. Other parameters are associated with atoms and are renormalized. With the renormalization factor Λ, these parameters are related to a set of bare parameters, U_0, U_1, g_0 and γ_0, via the standard renormalization relations [88],

$$U_a = \Lambda U_0, \quad U_{ab} = \Lambda U_1, \tag{5.128}$$

$$g = \Lambda g_0, \quad \gamma = \gamma_0 - (\Lambda g_0^2/U_c). \tag{5.129}$$

The renormalization factor is given by

$$\Lambda = (1 + (U_0/U_c))^{-1}, \quad U_c^{-1} = -\sum_k \frac{\mathrm{e}^{-k^2/k_c^2}}{2\varepsilon_k}, \tag{5.130}$$

with the cutoff momentum k_c representing the inverse range of interactions [104–106]. The bare parameters are

$$\gamma_0 = \mu_{co}(B - B_0), \quad g_0 = \sqrt{\frac{4\pi\hbar^2 a_{bg}\Delta B \mu_{co}}{m_a}}, \tag{5.131}$$

$$U_0 = \frac{4\pi\hbar^2 a_{bg}}{m_a}, \quad U_1 = \frac{4\pi\hbar^2 a_{ab}}{m_{ab}}, \tag{5.132}$$

where B is the applied magnetic field, which changes linearly with time at a rate of α_r, i.e., $B = -\alpha_r t$. B_0 and ΔB are the position and width, respectively, of the Feshbach resonance. m_a and $m_b = 2m_a$ are the masses of the atoms and molecules, respectively, and $m_{ab} = \frac{2}{3}m_a$ is the reduced mass of the atom-molecule interaction. In addition, μ_{co} is the difference in magnetic moment between the two channels, and we have assumed that the s-wave scattering length near resonance has the form $a_s = a_{bg}(1 - (\Delta B/(B - B_0)))$, with a_{bg} being the background atomic scattering length. The scattering lengths of the atom-molecule and molecule-molecule interactions are denoted by a_{ab} and a_{bb}, respectively.

Due to the trapping potential in experiments, the molecular bosons are more tightly confined in space than the fermionic atoms due to their different statistics [107]. To show this inequality, we use V_a for the volume of fermionic atoms and V_b for bosonic molecules. We consider the zero-temperature limit, where we can consider only one bosonic mode and ignore all possible dissipations in the system, such as the loss of atoms by three-body collisions.

Due to the presence of an external magnetic field, the 'spin-up' and 'spin-down' states actually have a Zeeman component (h) to their energy, i.e., $\varepsilon_{k\uparrow} = \varepsilon_k + h$, $\varepsilon_{k\downarrow} = \varepsilon_k - h$. However, the total energy of the non-interacting atoms $\sum_{k,\sigma} \varepsilon_{k,\sigma} a_{k,\sigma}^\dagger a_{k,\sigma}$ can be rewritten as

$$\sum_k \left[\varepsilon_k (a_{k\uparrow}^\dagger a_{k\uparrow} + a_{k\downarrow}^\dagger a_{k\downarrow}) + h(a_{k\uparrow}^\dagger a_{k\uparrow} - a_{k\downarrow}^\dagger a_{k\downarrow})\right]. \tag{5.133}$$

In our discussion, the numbers of 'spin-up' atoms and 'spin-down' atoms are the same. Therefore, the second term in the above expression vanishes, and we have not included the Zeeman energy term in Eq. (5.127).

In the current experiments, the intrinsic energy width of a Feshbach resonance is larger than the Fermi energy E_F [108]; it is therefore reasonable to assume $\varepsilon_k = \varepsilon$. This approximation is called the degenerate model in [85, 97, 103] and was verified by exact numerical calculations in [85, 97]. Here, we use this degenerate approximation.

We proceed by introducing the following operators [85, 97]:

$$L_x = \frac{\sum_k (a^\dagger_{k,\uparrow} a^\dagger_{-k,\downarrow} b + b^\dagger a_{-k,\downarrow} a_{k,\uparrow})}{(N/2)^{3/2}}, \tag{5.134}$$

$$L_y = \frac{\sum_k (a^\dagger_{k,\uparrow} a^\dagger_{-k,\downarrow} b - b^\dagger a_{-k,\downarrow} a_{k,\uparrow})}{(N/2)^{3/2}}, \tag{5.135}$$

$$L_z = \frac{\sum_{k,\sigma} a^\dagger_{k,\sigma} a_{k,\sigma} - 2b^\dagger b}{N}, \tag{5.136}$$

where $N = 2b^\dagger b + \sum_{k,\sigma} a^\dagger_{k,\sigma} a_{k,\sigma}$ is the total number of atoms. The Hamiltonian (5.127) becomes

$$\begin{aligned} H = &\frac{N}{4}\left[2\varepsilon - \left(\gamma + \frac{\varepsilon_b}{2}\right) - \frac{NU_a}{2V_a} - \frac{NU_{ab}}{V_a}\right] L_z \\ &- \frac{N^2}{16}\left(\frac{U_a}{V_a} + \frac{2U_{ab}}{V_a} - \frac{U_b}{V_b}\right)(1 - L_z)^2 + \frac{gV_b}{V_a^{3/2}}\left(\frac{N}{2}\right)^{3/2} L_x, \end{aligned} \tag{5.137}$$

with the commutators

$$[L_z, L_x] = \frac{4i}{N} L_y, \quad [L_z, L_y] = -\frac{4i}{N} L_x, \tag{5.138}$$

$$[L_x, L_y] = \frac{i}{N}(1 - L_z)(1 + 3L_z) + O(1/N^2). \tag{5.139}$$

In deducing the atom-atom scattering term, we need to introduce the collective pseudo-spin operators: $\hat{S}_z = \sum_k (a^\dagger_{k,\uparrow} a_{k,\uparrow} + a^\dagger_{-k,\downarrow} a_{-k,\downarrow} - 1)/2$, $\hat{S}^+ = \sum_k a^\dagger_{k,\uparrow} a^\dagger_{-k,\downarrow}$, and $\hat{S}^- = \sum_k a_{-k,\downarrow} a_{k,\uparrow}$. It is straightforward to prove that $\hat{S}^2 = \hat{S}_z^2 - \hat{S}_z + \hat{S}^+\hat{S}^-$ is a conservation law and that $S = N/4$. Combined with the conserved relation of the total particles, $N/4 = b^\dagger b + \hat{S}_z$, we can rewrite the atom-atom scattering term as $\hat{S}^+\hat{S}^- = (1/2)\sum_{k,\sigma} a^\dagger_{k,\sigma} a_{k,\sigma} b^\dagger b + (N/2) - b^\dagger b$.

We can obtain the Heisenberg equations for the system as follows:

$$\hbar\frac{dL_x}{dt} = -\left[2\varepsilon - \left(\gamma + \frac{\varepsilon_b}{2}\right) - \frac{NU_a}{2V_a} - \frac{NU_{ab}}{V_a}\right]L_y$$
$$-\frac{N}{4}\left(\frac{U_a}{V_a} + \frac{2U_{ab}}{V_a} - \frac{U_b}{V_b}\right)[(1-L_z)L_y + L_y(1-L_z)], \quad (5.140)$$
$$\hbar\frac{dL_y}{dt} = \left[2\varepsilon - \left(\gamma + \frac{\varepsilon_b}{2}\right) - \frac{NU_a}{2V_a} - \frac{NU_{ab}}{V_a}\right]L_x$$
$$+\frac{N}{4}\left(\frac{U_a}{V_a} + \frac{2U_{ab}}{V_a} - \frac{U_b}{V_b}\right)[(1-L_z)L_x + L_x(1-L_z)]$$
$$-\frac{\sqrt{2N}gV_b}{4V_a^{3/2}}(1-L_z)(1+3L_z) + O(1/\sqrt{N}), \quad (5.141)$$
$$\hbar\frac{dL_z}{dt} = \frac{\sqrt{2N}gV_b}{V_a^{3/2}}L_y. \quad (5.142)$$

In the mean-field approximation, we replace the operators in the above equations with their expectations, such as using $\langle L_x\rangle$ for L_x. However, these equations show that the expectation values of the single operators, e.g., $\langle L_x\rangle$, depend not only on themselves but also on the second-order moments, e.g., $\langle L_xL_y\rangle$. Similarly, the time evolution of the second-order moments depends on the third-order moments, and so on. Consequently, we obtain a hierarchy of equations of motion for the expectation values. To obtain a closed set of equations of motion, the hierarchy must be truncated at some stage by approximating the Nth-order moments in terms of lower-order moments [85, 109]. The lowest-order truncation is achieved by approximating the second-order moments with the products of the expectation values of the corresponding single operators, such as $\langle L_xL_y\rangle$ with $\langle L_x\rangle \cdot \langle L_y\rangle$. This truncation is further justified by the following fact. In our discussion, the total number of atoms N is large. We note that the commutators in Eqs. (5.138) and (5.139) vanish and that L_x, L_y and L_z commute with each other in the limit of large N. In this case, one typically applies factorization relations such as $\langle L_xL_y\rangle = \langle L_x\rangle \cdot \langle L_y\rangle$ [110].

With the introduction of the three real numbers u, v, and w for the expectation values of the three operators L_x, L_y, and L_z, respectively, and ignoring the $O(1/\sqrt{N})$ terms, the above Heisenberg equations become a set of mean-field equations:

$$\frac{du}{d\tau} = -\delta v - 2\chi v(1-w), \quad \frac{dw}{d\tau} = \sqrt{2}v, \quad (5.143)$$
$$\frac{dv}{d\tau} = \frac{3\sqrt{2}}{4}(w-1)(w+\frac{1}{3}) + \delta u + 2\chi u(1-w), \quad (5.144)$$

where

$$\tau = \frac{\sqrt{N} g V_b}{\hbar V_a^{3/2}} t, \tag{5.145}$$

$$\delta = \left[2\varepsilon - \left(\gamma + \frac{\varepsilon_b}{2} \right) - \frac{N U_a}{2 V_a} - \frac{N U_{ab}}{V_a} \right] \frac{V_a^{3/2}}{\sqrt{N} g V_b}, \tag{5.146}$$

$$\chi = \left(U_a + 2U_{ab} - \frac{U_b V_a}{V_b} \right) \frac{\sqrt{N V_a}}{4 g V_b}. \tag{5.147}$$

Because of the identity $u^2 + v^2 = \frac{1}{2}(w-1)^2(w+1)$, only two independent variables exist. By introducing the variable $\theta = \arctan(v/u)$, which is canonically conjugate to w, we obtain the classical Hamiltonian:

$$\mathscr{H} = \delta w - \chi(1-w)^2 + \sqrt{(w-1)^2(w+1)} \cos\theta. \tag{5.148}$$

These equations show that all the experimental parameters affect the system via only two dimensionless parameters: δ and χ. By a trivial shift of the time origin, we can set $\delta = \alpha\tau$, with

$$\frac{\alpha_r}{\alpha} = \frac{4\pi \hbar n a_{bg} \Delta B}{m_a} \Lambda^2 \frac{V_b^2}{V_a^2}, \tag{5.149}$$

where $n = N/V_a$ is the mean atomic density, α is the scaled sweep rate, and τ is the scaled time. The nonlinear parameter χ is given by

$$\chi = \frac{1}{2} \left(1 + \frac{3a_{ab}}{a_{bg}} - \frac{a_{bb} V_a}{2 a_{bg} \Lambda V_b} \right) \frac{V_a}{V_b} \sqrt{\frac{\pi \hbar^2 a_{bg} n}{m_a \mu_{co} \Delta B}}. \tag{5.150}$$

The Hamiltonian (5.148) has the energy unit of $\frac{4V_a^{3/2}}{g V_b N^{3/2}}$. The variable w measures the imbalance between atom pairs and molecules and varies in the range of $[-1, 1]$, with $w = -1$ corresponding to a pure molecular gas and $w = 1$ to a pure atomic gas. We are interested in how many atomic pairs are converted to molecules after the magnetic field crosses the resonance. We use w_f to denote the value of w long after the magnetic field has passed the resonance. The molecular conversion efficiency is defined as $T = 1 - \Gamma = (1 - w_f)/2$, while the fraction of unconverted atoms is defined as $\Gamma = (1 + w_f)/2$.

To understand the dynamics of the Hamiltonian (5.148), we first look at the fixed points of this system. These points can be found by setting $dw/dt = du/dt = dv/dt = 0$ in (5.143) and (5.144). The energies for these fixed points make up energy levels of the system as shown in Fig. 5.33. We find that the structure of these energy levels changes dramatically as the nonlinear parameter χ increases.

Consider the adiabatic evolution of the system starting from a high negative value of δ with $w = 1$. This condition corresponds to experiments in which the magnetic field sweeps slowly across the Feshbach resonance with initially no bosonic

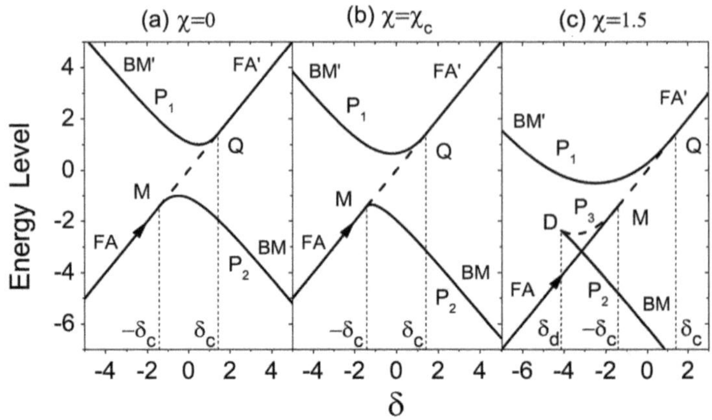

Fig. 5.33 Adiabatic energy levels for different interaction strengths. (a) $\chi = 0$; (b) $\chi = \chi_c = \sqrt{2}/4$; and (c) $\chi = 1.5$. The unstable states are indicated by dashed lines (MQ and DM)

molecules. When χ is small, as in Fig. 5.33a, the evolution of the system follows the solid line, converting all fermionic atoms into molecules. However, when χ exceeds $\chi_c = \sqrt{2}/4$, as in Fig. 5.33c, the system finds no stable energy level to follow at a single point M. As a result, only a fraction of the fermionic atoms are converted into bosonic molecules.

This simple analysis is confirmed by our numerical results, which are plotted in Fig. 5.34. Because $w = 1$ is a fixed point when $\delta < -\sqrt{2}$, we start from $(w, u, v) \approx (1, 0, 0)$ and sweep the field from $\delta = -\sqrt{2}$ to 200. Then, w_f is recorded, and the conversion efficiency T obtained by using the relation $T = (1 - w_f)/2$. In Fig. 5.34, the conversion efficiency T, i.e., the fraction of the converted fermionic atom pairs, is drawn as a function of α. T approaches 1 as $\alpha \to 0$ when $\chi < \chi_c$, indicating that all atomic pairs are converted into molecules. In contrast, when $\chi > \chi_c$, T does not increase to 1 in the adiabatic limit $\alpha \to 0$. Consequently, there is an upper limit T_{ad} (<100%) to the conversion efficiency. Moreover, Fig. 5.34 demonstrates that positive χ suppresses the conversion efficiency, whereas negative χ enhances it. Because the repulsive interaction between bosonic molecules enters χ as a negative value, this interaction enhances the conversion efficiency; the repulsive fermionic atomic interaction and atom-molecule interaction contribute positively to χ and thus suppress the conversion.

The upper limit T_{ad} on the atom-molecule conversion efficiency depends on χ. This dependence can be found by examining the phase-space diagrams of our system, as shown in Fig. 5.35. As δ increases slowly from a large negative value, the fixed point P_3 moves up until it intersects the fixed point ($w = 1, u = 0, v = 0$), represented by a dark straight line in Fig. 5.35a. This collision occurs at $\delta = -\sqrt{2}$. Immediately after the collision, the hyperbolic fixed point P_3 is no longer a fixed point and becomes a solution that evolves along the dark line in Fig. 5.35b. The dark line is given by $\sqrt{2} = \chi(1 - w) - \sqrt{1 + w}\cos\theta$, which is found by taking $E = \delta = -\sqrt{2}$ in (5.148).

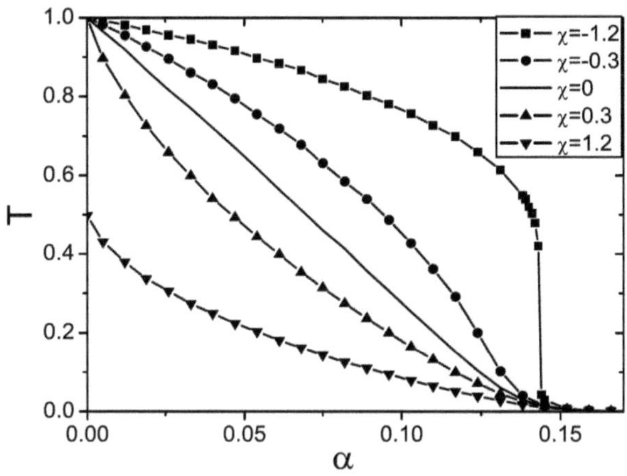

Fig. 5.34 Conversion efficiency T as a function of the sweep rate α for various interactions

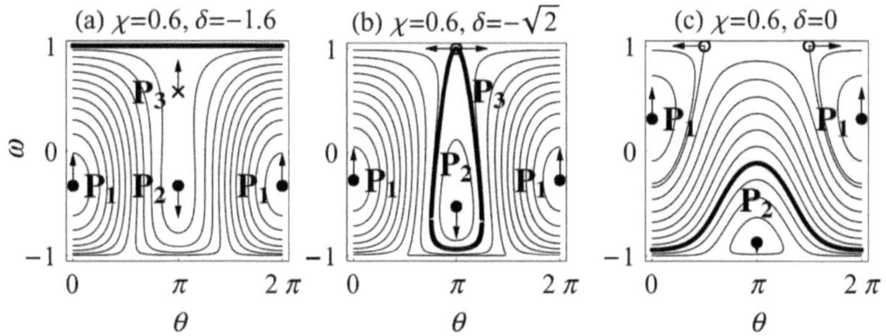

Fig. 5.35 Phase spaces of the Hamiltonian (5.148). The dark line in (**a**) represents the fixed point $w = 1$, $u = 0$, $v = 0$. It is a line because θ is not defined at $u = v = 0$. The two fixed points on line $w = 1$ in (**b**) are in fact the same fixed point; they are an artifact caused by the definition $\theta = \arctan(v/u)$

As the action of this trajectory is nonzero, whereas a fixed point has zero action, this collision of the two fixed points represents a sudden jump in action. This sudden jump results in a nonzero fraction of remnant atoms. As δ continues to increase slowly, the trajectory changes its shape, as shown in Fig. 5.35c; however, its action stays constant, as required by the classical adiabatic theorem [4, 33]. The action is

$$I = \begin{cases} \frac{1}{2\pi}\oint \frac{\cos\theta\sqrt{8\chi^2-4\sqrt{2}\chi+\cos^2\theta}}{2\chi^2}d\theta, & \frac{\sqrt{2}}{4} < \chi < \frac{\sqrt{2}}{2}; \\ \frac{1}{2\pi}\int_0^{2\pi} \frac{4\chi^2-2\sqrt{2}\chi+\cos^2\theta}{2\chi^2}d\theta, & \chi > \frac{\sqrt{2}}{2}. \end{cases} \tag{5.151}$$

According to the definition of the action, we have the relation $I = w_f + 1$. Using the relation between the conversion efficiency and the variable w_f, we obtain an upper limit on the efficiency in the adiabatic limit [86],

$$T_{ad} = \begin{cases} \frac{4\sqrt{2}\chi - 1}{8\chi^2}, & \chi > \frac{\sqrt{2}}{4}; \\ 1, & \chi < \frac{\sqrt{2}}{4}. \end{cases} \quad (5.152)$$

5.6 Nonlinear Composite Adiabatic Passage

Manipulating the state of a quantum system by external fields is crucial in atomic and molecular physics for applications such as metrology, interferometry, nuclear magnetic resonance, quantum information processing, and the driving of chemical reactions [111–114]. The practical implementation of quantum information processing, however, requires time-dependent schemes featuring three important issues: The driving quantum state to a target state should be achieved (i) with high fidelity, typically with an admissible error less than 10^{-4}, (ii) in the shortest possible time to prominently minimize decoherence effects, and (iii) in a robust way with respect to the imperfect knowledge of the system or to variations in experimental parameters [115–117].

The adiabatic passage (AP) techniques are a widely used tool for quantum state manipulation. A variety of AP techniques have been proposed and demonstrated, including rapid adiabatic passage, Stark-chirped rapid adiabatic passage, piecewise adiabatic passage, and stimulated Raman adiabatic passage and its variations [118]. The techniques are robust, but in nearly all of them, the transition probability is incomplete. Another basic approach to robust coherent control of quantum systems is the technique of composite pulses, which is widely used in nuclear magnetic resonance [119], in quantum optics, and in quantum information processing [120, 121]. This technique replaces the single pulse used conventionally for driving a two-state transition by a sequence of pulses with appropriately chosen phases, which are used as a control tool for shaping the excitation profile in a desired manner, e.g., to make the system more robust to variations in the experimental parameters-intensities and frequencies. The imperfections may be caused by an imprecise pulse area, an undesirable frequency offset, or an unwanted frequency chirp [122].

To combine the advantages of adiabatic passage and composite pulse techniques and to achieve robust and high-fidelity quantum state control, multiple optimal control approaches have been proposed [123–126]. Among them, the composite adiabatic passage (CAP) technique is a powerful and flexible control tool [126] that can deliver an extremely high fidelity of population transfer, far beyond the fault-tolerant quantum computing benchmark. The CAP technique has been widely studied [127] and demonstrated experimentally in a rare-earth ion-doped solid [128]. The experimental explanation and associated theoretical discussion are limited to a linear two-level system in which the interaction between particles is ignored. However, interest in studying the nonlinear quantum system with interparticle interaction has been increasing. The interaction between the particles can significantly influence the quantum transition dynamics [4, 129]. Moreover, the single-pulse duration must be very long (i.e., an infinitely slow sweep speed) to satisfy the adiabaticity criteria.

However, a CAP sequence contains N pulses. Hence, the total pulse duration for CAP is N times the single-pulse duration. Under adiabatic conditions, the total pulse duration is infinite, which is impractical (and indeed, unphysical).

In this section, we discuss the CAP and achieve high-fidelity, fast, and robust quantum state manipulation in a nonlinear two-level system in which the level energies depend on the occupation of the levels, representing the mean-field interaction between the particles. The influence of interparticle interaction on the CAP is demonstrated. We show that the interaction tends to increase the number of pulse sequences, i.e., the high-fidelity transition probability can still be achieved in a nonlinear system as long as there exist sufficiently long composite sequences. Different from the linear quantum system, regardless of the number of pulse sequence, the total pulse duration is fixed for nonlinear quantum systems [130].

The nonlinear two-state system we consider is described by the following dimensionless Schrödinger equation [131]:

$$i\frac{\partial}{\partial t}\boldsymbol{a}(t) = \boldsymbol{H}(t)\boldsymbol{a}(t), \tag{5.153}$$

with the Hamiltonian given by

$$\boldsymbol{H}(t) = \frac{\nu(t)}{2}\hat{\sigma}_x + \left[\frac{\gamma(t)}{2} + \frac{c}{2}(|a_2|^2 - |a_1|^2)\right]\hat{\sigma}_z, \tag{5.154}$$

where $\boldsymbol{a}(t) = [a_1(t), a_2(t)]^T$ is the probability amplitudes of the two states. $\hat{\sigma}_x$ and $\hat{\sigma}_z$ are the Pauli matrices. $\gamma(t)$ and $\nu(t)$ are the energy bias and coupling strength between the two states, respectively. c is the nonlinear parameter describing the interparticle interaction. The total probability $|a_1|^2 + |a_2|^2$ is conserved and set to 1. The model has aroused great interest in the context of theory and has important applications in physics, for example, for describing the spin tunneling of nanomagnets [132], a BEC in a double-well potential or in an optical lattice [133], and coupled waveguide arrays [134].

In the linear model, i.e., $c = 0$, the CAP method has been proposed [126], in which the propagator of the two-state system can be parameterized by the Cayley-Klein parameters and the single pulse driving the quantum transition is replaced by a sequence of pulses with appropriately chosen phases. The technique allows one to suppress the nonadiabatic oscillations in the transition probability and to reduce the error below the 10^{-4} quantum computation benchmark, even with simple three- and five-pulse composite sequences. In addition, the composite phases do not depend on the specific pulse shape and chirp as long as the latter satisfies the symmetry property.

The success of the CAP in linear systems has demonstrated its strong ability to realize quantum manipulation [127, 128]. Keeping this in mind, for the system (5.153) with Hamiltonian (5.154), we employ a sequence of N ($N = 2n + 1$, n is an integer) pulses, each with a phase ϕ_k ($k = 1, 2, \cdots, N$), to achieve high-fidelity quantum transition. The phase ϕ_k is imposed upon the driving field Rabi frequency (coupling strength), $\nu(t) \rightarrow \nu(t)e^{i\phi_k}$. For simplicity and as a first attempt toward a

CAP protocol for nonlinear systems, the composite control phase in the linear systems [126] is used here. Assuming that the coupling strength $v(t)$ is an even function of time and that the detuning $\gamma(t)$ is odd, the composite phase is

$$\phi_k = \left(N + 1 - 2\left\lfloor \frac{k+1}{2} \right\rfloor\right) \left\lfloor \frac{k}{2} \right\rfloor \frac{\pi}{N}, \tag{5.155}$$

where the symbol $\lfloor x \rfloor$ denotes the floor function. The phase sequence is symmetric, i.e., $\phi_k = \phi_{N+1-k}$, and $\phi_1 = \phi_N = 0$.

Our goal is to show the CAP with nonlinear interparticle interaction and to consider how the nonlinear interaction affects the CAP technique. With the emergence of nonlinearity, the transition dynamics change dramatically. In this case, the Schrödinger equation (5.153) is no longer analytically solvable. We therefore exploit a 4–5th-order Runge-Kutta algorithm to trace the quantum evolution numerically and to calculate the transition probability of the system.

As an example, we consider the Allen-Eberly (AE) model assuming a sech-type coupling strength (pulse) and a tanh-type frequency energy bias (chirp) [126],

$$v(t) = v_0 \mathrm{sech}(t/T), \quad \gamma(t) = \alpha \tanh(t/T), \tag{5.156}$$

where v_0 and α are constant parameters with the dimension of frequency and T is the pulse width.

For the linear case, the transition probability $p = |a_2|^2 = 1 - |a_1|^2$ is [126]

$$\begin{aligned} p &= \frac{\cosh(\pi\alpha T) - \cos(\pi T\sqrt{v_0^2 - \alpha^2})}{1 + \cosh(\pi v_0 T)} \\ &= 1 - \frac{\cos^2(\frac{1}{2}\pi T\sqrt{v_0^2 - \alpha^2})}{\cosh^2(\frac{\pi\alpha T}{2})}. \end{aligned} \tag{5.157}$$

For $v_0 < \alpha$, the cosine in (5.157) is replaced by a hyperbolic cosine. A transition probability $p = 1$ (complete population inversion) is obtained for $\sqrt{v_0^2 - \alpha^2 T} = 2n + 1$, with $n = 0, 1, 2, \cdots$ (integer). In the adiabatic limit ($v_0 > \alpha \gg 2/T$), the transition probability also tends to unity. If α is not large enough, nonadiabatic oscillations mediated by v_0 appear, and the probability is reduced. These oscillations can be suppressed to any order by the CAP technique even with simple three- and five-pulse composite sequences. Note also that all the variables here should be considered scaled dimensionless variables. Throughout, we use T for scaling; thus, $T = 1$, and v_0, α, and c are in units of $1/T$.

Figure 5.36 shows the final transition probability as a function of both the number N of composite sequence pulses and the particle interaction c for the AE model with $\alpha = 1$ and $v_0 = 1.2$. The blue zones correspond to a low transition probability, whereas red areas indicate a high transition probability. The transition dynamics are

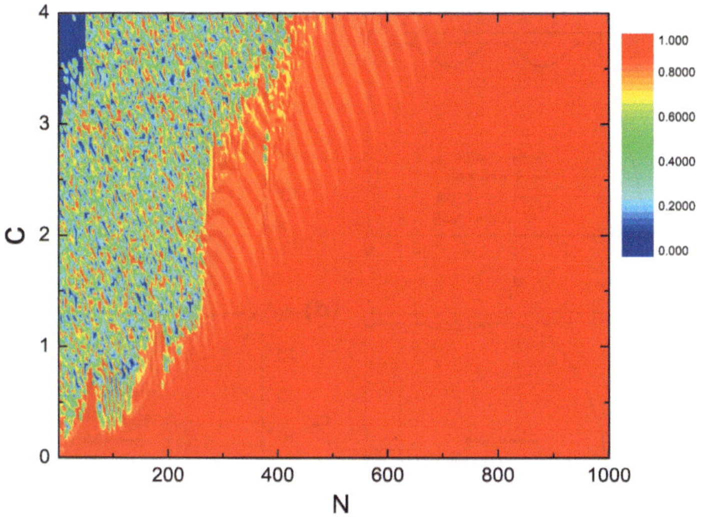

Fig. 5.36 (Color online) Contour plots of the transition probability as a function of composite sequence pulses N and interaction c for an AE pulse with $\alpha = 1$ and $\nu_0 = 1.2$

strongly dependent on the nonlinear interaction. For very weak interaction, a high transition probability can be achieved even with simple three- and five-pulse composite sequences. As the nonlinear interaction grows, nonadiabatic oscillations are significantly strengthened, and the probability is reduced dramatically. Consequently, the high transition probability can no longer be achieved with a small number of composite sequences. Interestingly, however, these oscillations can be suppressed by the CAP technique with sufficiently long composite sequences; i.e., as long as there exist sufficiently long composite sequences, the CAP technique can still be applied to the nonlinear two-level system. Note that for linear systems, the total pulse duration of the CAP is N times the single-pulse duration. However, the total pulse duration is fixed in our nonlinear system. In all numerical simulations, the numerical time was performed from times -100 to 100.

We calculate the transition probability as a function of the peak Rabi frequency ν_0 for different interaction and composite sequence pulses N; the results are shown in Fig. 5.37. Frames (a) and (b) show that a 1299-pulse CAP with interaction $c = 0.2$ suffices to suppress the nonadiabatic oscillations below the quantum information benchmark 10^{-4}. Frames (c) and (d) depict the transition probability versus the peak Rabi frequency ν_0 for interaction $c = 2.0$. For comparison with [126], we also plot the transition probability as a function of the peak Rabi frequency ν_0 for a five-pulse sequences $N = 5$ with $c = 0$ (olive dashed curve). For the linear system, even a five-pulse sequence is sufficient to achieve extremely high fidelity with an error below 10^{-4}. For nonlinear systems, the 10^{-4} error benchmark can still be reached, albeit with longer sequences.

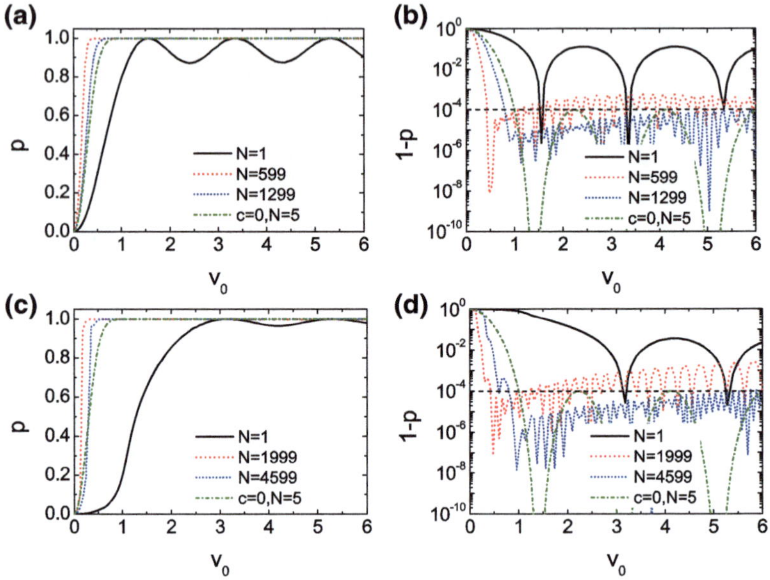

Fig. 5.37 (Color online) Transition probability as a function of the peak Rabi frequency v_0 for different interactions [$c = 0.2$ (top) and $c = 2.0$ (bottom)] and composite sequence pulses N. Frames (b) and (d) show the deviations of the left profiles. The olive dashed curve represents the five-pulse sequences $N = 5$ with $c = 0$. In addition, $\alpha = 1$

To test the robustness against variations in the field parameters, we vary both α and v_0 around their optimal values and calculate the fidelity. The results are summarized in Fig. 5.38, which clearly shows that CAP is extremely robust with respect to an increase in α and v_0. Furthermore, the high-fidelity region with an error below 10^{-4} for CAP is considerably expanded relative to a single pulse. Different from the linear quantum system, a large number of composite pulses is needed to achieve ultrahigh fidelity in CAP in nonlinear quantum systems, and the number of pulses gradually increases with increasing nonlinear interaction.

To further explore this peculiar phenomena, we introduce the relative phase $\theta = \theta_2 - \theta_1$ and the transition probability $p = |a_2|^2$ as two canonical conjugate variables, with $a_1 = \sqrt{1-p}e^{(i\theta_1)}$ and $a_2 = \sqrt{p}e^{(i\theta_2)}$. We can then obtain an effective classical Hamiltonian that satisfies the canonical equations, i.e., $dp/dt = -\partial H/\partial\theta$, $d\theta/dt = \partial H/\partial p$, and

$$H(t) = \frac{\gamma}{2}(1-2p) - \frac{c}{4}(1-2p)^2 + v\sqrt{p(1-p)}\cos(\theta + \phi_k). \qquad (5.158)$$

The classical Hamiltonian can describe completely the dynamic properties of the system (5.153) [4]. In Fig. 5.39, we show the transition probability and relative phase evolution for the AE model during CAP with and without the control phase under different interaction and pulse sequences. In Fig. 5.39a–c, we plot the trajectories

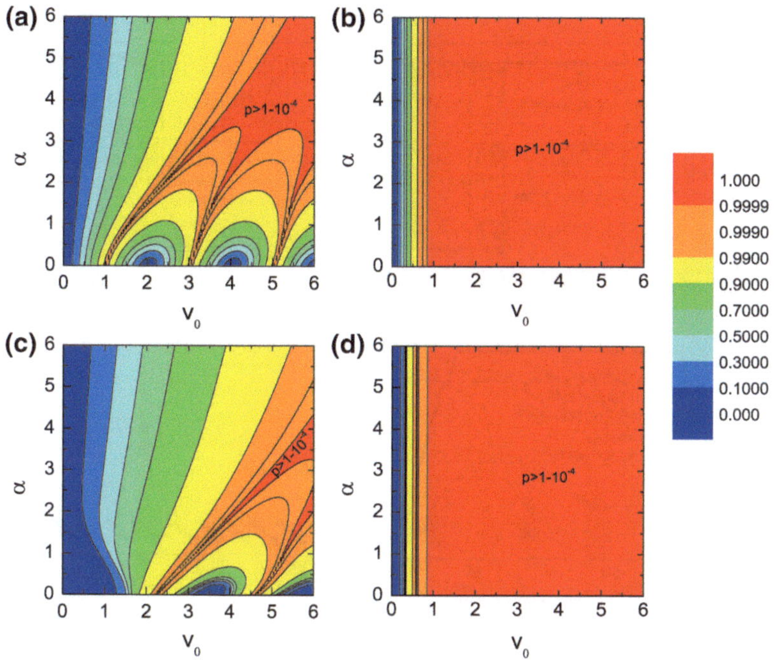

Fig. 5.38 (Color online) Contour plots of the transition probability as a function of the peak Rabi frequency ν_0 and chirp rate α at different interactions [$c = 0.2$ (top) and $c = 2.0$ (bottom)] with composite sequence pulses. **a** $N = 1$, **b** $N = 1399$, **c** $N = 1$, and **d** $N = 4799$

in the phase space, the transition probability and the relative phase evolution for various parameter values. The black triangle and the star represent the initial state and the final state, respectively. The control phases play an important role in the CAP technique and can significantly influence the quantum transition dynamics. This process allows one to suppress the nonadiabatic oscillations in the transition probability and to ensure an ultrahigh fidelity in the CAP process. In the linear case, the CAP works for a small number of pulses, and each constituent pulse produces a large population change but not complete inversion; the destructive interference of the deviations ultimately drives the system to complete inversion. In contrast, in the nonlinear quantum system, the CAP requires a large number of pulses, each of which produces a small change in population. However, the universal composite phases are derived from the condition that cancels the deviations from unit transfer efficiency due to nonadiabatic effects by enforcing destructive interference of these deviations. By directly solving the Schrödinger equation (5.153) using the same approach, we can reproduce the above results. In contrast to the CAP technique of [126], regardless of the number of pulse sequence, the total pulse duration is fixed. Thus, the profiles of the pulse and chirp are different for different pulse sequences. For example, for

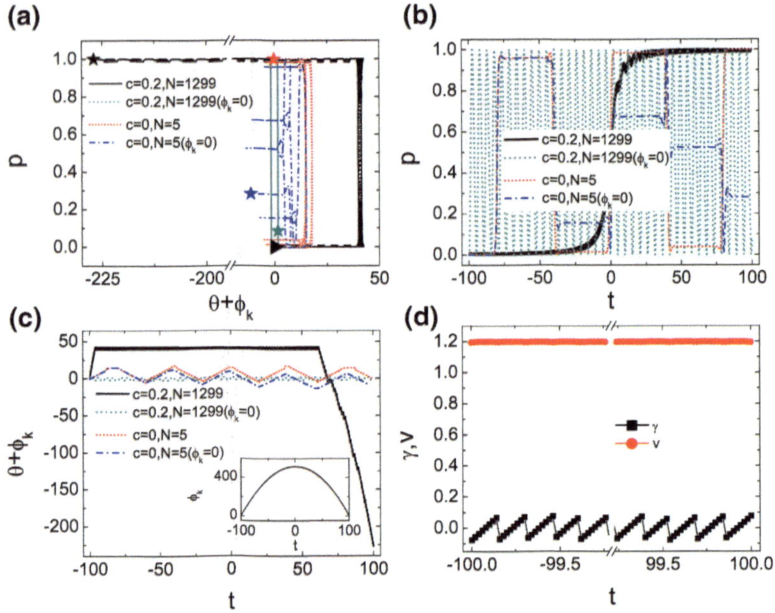

Fig. 5.39 (Color online) **a** Trajectories in the phase space. **b** Transition probability as a function of time (inset: control phase as a function of time). **c** Relative phase as a function of time. **d** Sequences of the 1299 pulse as a function of time

the 1299 pulse, the AE model reduces to the Landau-Zener model of finite duration [see Fig. 5.39d], which causes the transition probability to exhibit oscillations without the control phase.

References

1. Milburn, G.J., Corney, J., Wright, E.M., Walls, D.F.: Phys. Rev. A **55**, 4318 (1997); Vardi, A., Anglin, J.R.: Phys. Rev. Lett. **86**, 568 (2001); Kohler, S., Sols, F.: Phys. Rev. Lett. **89**, 060403 (2002)
2. Wu, B., Niu, Q.: Phys. Rev. A **61**, 023402 (2000)
3. Liu, J., Fu, L.B., Ou, B.Y., Chen, S.G., Choi, D.I., Wu, B., Niu, Q.: Phys. Rev. A **66**, 023404 (2002)
4. Liu, J., Wu, B., Niu, Q.: Phys. Rev. Lett. **90**, 170404 (2003)
5. Pu, H., Maenner, P., Zhang, W., Ling, H.Y.: Phys. Rev. Lett. **98**, 050406 (2007)
6. Liu, J., Fu, L.B.: Phys. Rev. A **81**, 052112 (2010)
7. Liu, J., Fu, L.B.: Ann. Phys. (N.Y.) **325**, 2425 (2010)
8. Lahini, Y., Pozzi, F., Sorel, M., Morandotti, R., Christodoulides, D.N., Silberberg, Y.: Phys. Rev. Lett. **101**, 193901 (2008)
9. Zhang, L.D., Fu, L.B., Liu, J.: Eur. Phys. J. D **65**, 557 (2011)
10. Jensen, S.M.: IEEE J. Quantum Electron. QE- **18**, 1580 (1982)
11. Agrawal, G.P.: Applications of Nonlinear Fiber Optics. Academic, San Diego (2001)

12. Yariv, A., Yue, P.: Photonics: Optical Electronics in Modern Communications. Oxford University, New York (2006)
13. Lederer, F., Darmanyan, S., Kobyakov, A.: Discrete solitons. In: Trillo, S., Torruellas, W.E. (eds.) Spatial Optical Solitons. Springer, New York (2001)
14. Coutinho dos Santos, B., Souza, C.E.R., Dechoum, K., Khoury, A.Z.: Phys. Rev. A **76**, 053821 (2007); Coutinho dos Santos, B., Dechoum, K., Khoury, A.Z.: Phys. Rev. Lett. **103**, 230503 (2009)
15. Mukunda, N., Simon, R.: Ann. Phys. (N.Y.) **228**, 205 (1993)
16. Berry, M.V.: Proc. Roy. Soc. A **392**, 45 (1984)
17. Wilczek, F., Zee, A.: Phys. Rev. Lett. **52**, 2111 (1984); Aharonov, Y., Anandan, J.: Phys. Rev. Lett. **58**, 1593 (1987); Samuel, J., Bhandari, R.: Phys. Rev. Lett. **60**, 2339 (1988)
18. Shapere, A., Wilczek, F.: Geometric Phase in Physics. World Scientific, Singapore (1989); Bohm, A., Mostafazadeh, A., Koizumi, H., Niu, Q., Zwanziger, J.: The Geometric Phase in Quantum Systems. Springer, New York (2003)
19. Landau, L.D., Lifshitz, E.M., Pitaevskii, E.M.: Statistical Physics. Butterworth-Heinemann, New York (1999)
20. Carollo, A.C.M., Pachos, J.K.: Phys. Rev. Lett. **95**, 157203 (2005); Zhu, S.L.: Phys. Rev. Lett. **96**, 077206 (2006); Hamma, A.: quant-ph/0602091 (2006); Plastina, F., Liberti, G., Carollo, A.: Europhys. Lett. **76**, 182 (2006); Chen, G., Li, J., Liang, J.Q.: Phys. Rev. A **74**, 054101 (2006); Cui, H.T., Li, K., Yi, X.X.: Phys. Lett. A **360**, 243 (2006); Zhang L.D., Fu, L.B. Europhys. Lett. **93**, 30001 (2011)
21. Grifoni, M., Hanggi, P.: Phys. Rep. **304**, 229 (1998)
22. Landau, L.D.: Phys. Z. Sowjetunion 2, **46** (1932); Zener, G.: Proc. R. Soc. Lond. Ser. A **137**, 696 (1932)
23. Mullen, K., Ben-Jacob, E., Schuss, Z.: Phys. Rev. Lett. **60**, 1097 (1988); Mullen, K., et al.: Physica B **153**, 172 (1988); Greigenmüller, U., Schön, G.: Physica B **153**, 186 (1988)
24. Bharucha, C.F., Madison, K.W., Morrow, P.R., Wilkinson, S.R., Sundaram, B., Raizen, M.G.: Phys. Rev. A **55**, R857 (1997); Madison, K.W., Bharucha, C.F., Morrow, P.R., Wilkinson, S.R., Niu, Q., Sundaram, B., Raizen, M.G.: Appl. Phys. B Lasers Opt. B **65**, 693 (1997)
25. Sibille, A., Palmier, J.F., Laruelle, F.: Phys. Rev. Lett. **80**, 4506 (1998)
26. Eilbeck, J.C., Lomdahl, P.S., Scott, A.C.: Phys. D **16**, 318 (1985); Kenkre, V.M., Campbell, D.K.: Phys. Rev. B **34**, 4959 (1986); Datta, P.K., Kundu, K. Phys. Rev. B **53**, 14929 (1996)
27. Andrews, M.R., Townsend, C.G., Miesner, H.J., Durfee, D.S., Kurn, D.M., Ketterle, W.: Science **275**, 637 (1997)
28. Smerzi, A., Fantoni, S., Giovanazzi, S., Shenoy, S.R.: Phys. Rev. Lett. **79**, 4950 (1997); Milburn, G.J., et al.: Phys. Rev. A **55**, 4318 (1997)
29. Mewes, M.O., Andrews, M.R., Kurn, D.M., Durfee, D.S., Townsend, C.G., Ketterle, W.: Phys. Rev. Lett. **78**, 582 (1997)
30. Anderson, M.H., Matthews, M.R., Wieman, C.E., Cornell, E.A.: Science **269**, 198 (1995)
31. Choi, D., Niu, Q.: Phys. Rev. Lett. **82**, 2022 (1999)
32. Zobay, O., Garraway, B.M.: Phys. Rev. A **61**, 033603 (2000)
33. Landau, L.D., Lifshitz, E.M.: Mechanics Pergamon. Oxford (1977)
34. Carroll, C.E., Hioe, F.T.: J. Phys. A Math. Gen. **19**, 2061 (1986)
35. Graefe, E.M., Korsch, H.J., Witthaut, D.: Phys. Rev. A **73**, 013617 (2006)
36. Yang, L.Y., Fu, L.B., Liu, J.: Laser Phys. **19**, 678 (2008)
37. Ostrovskaya, E.A., Kivshar, Y.S., Lisak, M., Hall, B., Cattani, F., Anderson, D.: Phys. Rev. A **61**, 031601(R) (1999)
38. Ananikian, D., Bergeman, T.: Phys. Rev. A **73**, 013604 (2006)
39. Rosen, N., Zener, C.: Phys. Rev. **40**, 502 (1932)
40. Rabi, I.I.: Phys. Rev. **51**, 652 (1937)
41. Thomas, G.F.: Phys. Rev. A **27**, 2744 (1983)
42. Osherov, V.I., Voronin, A.I.: Phys. Rev. A **49**, 265 (1994)
43. Kyoseva, E.S., Vitanov, N.V.: Phys. Rev. A **73**, 023420 (2006)
44. Olson, R.E.: Phys. Rev. A **6**, 1822 (1972)

45. Suominen, K.A., Garraway, B.M., Stenholm, S.: Phys. Rev. A **45**, 3060 (1992)
46. Bloch, F.: Phys. Rev. **70**, 460 (1946); Bloch, F., Hansen, W.W., Packard, M. Phys. Rev. **70**, 474 (1946)
47. Nielsen, M.A., Chuang, I.L.: Quantum Computation and Quantum Information. Cambridge Unversity Press, Cambridge (2000)
48. Ye, D.F., Fu, L.B., Liu, J.: Phys. Rev. A **77**, 013402 (2008)
49. Wang, G.F., Ye, D.F., Fu, L.B., Chen, X.Z., Liu, J.: Phys. Rev. A **74**, 033414 (2006)
50. Ramsey, N.F.: Phys. Rev. **78**, 695 (1950)
51. Mousavi, S.V., del Campo, A., Lizuain, I., Muga, J.G.: Phys. Rev. A **76**, 033607 (2007)
52. Seidel, D., Muga, J.G.: Phys. Rev. A **75**, 023811 (2007)
53. Santarelli, G., Laurent, P., Lemonde, P., et al.: Phys. Rev. Lett. **82**, 4619 (1999)
54. Fertig, C., Gibble, K.: Phys. Rev. Lett. **85**, 1622 (2000)
55. Gustavson, T.L., Bouyer, P., Kasevich, M.A.: Phys. Rev. Lett. **78**, 2046 (1997)
56. Dubetsky, B., Kasevich, M.A.: Phys. Rev. A **74**, 023615 (2006)
57. Peters, A., et al.: Nature (London) **400**, 849 (1999)
58. Weiss, D.S., Young, B.C., Chu, S.: Phys. Rev. Lett. **70**, 2706 (1993)
59. Weel, M., Kumarakrishnan, A.: Phys. Rev. A **67**, 061602(R) (2003)
60. Widera, A., et al.: Phys. Rev. Lett. **92**, 160406 (2004)
61. Davis, K.B.: Phys. Rev. Lett. **75**, 3969 (1995); Bradley, C.C., et al.: Phys. Rev. Lett. **75**, 1687 (1995)
62. Donley, E.A., Claussen, N.R., Thompson, S.T., Wieman, C.E.: Nature (London) **417**, 529 (2002)
63. Claussen, N.R., et al.: Phys. Rev. A **67**, 060701(R) (2003)
64. Góral, K., Köhler, T., Burnett, K.: Phys. Rev. A **71**, 023603 (2005)
65. Schumm, T., et al.: Nat. Phys. **1**, 57 (2005)
66. Lee, C.H.: Phys. Rev. Lett. **97**, 150402 (2006)
67. Li, S.C., Fu, L.B., Duan, W.S., Liu, J.: Phys. Rev. A **78**, 063621 (2008)
68. Leggett, A.J.: Rev. Mod. Phys. **73**, 307 (2001)
69. Yi, S., Müstecaplıoğlu, Ö.E., Sun, C.P., You, L.: Phys. Rev. A **66**, 011601(R) (2002)
70. Fu, L.B., Chen, S.G.: Phys. Rev. E **71**, 016607 (2005)
71. Hall, B.V., Whitlock, S., Anderson, R., Hannaford, P., Sidorov, A.I.: Phys. Rev. Lett. **98**, 030402 (2007)
72. Inouye, S., et al.: Nature (London) **392**, 151 (1998)
73. Heinzen, D.J., Wynar, R., Drummond, P.D., Kheruntsyan, K.V.: Phys. Rev. Lett. **84**, 5029 (2000)
74. Timmermans, E., et al.: Phys. Rep. **315**, 199 (1999)
75. Köhler, T., et al.: Rev. Mod. Phys. **78**, 1311 (2006)
76. Mies, F.H., Tiesinga, E., Julienne, P.S.: Phys. Rev. A **61**, 022721 (2000)
77. Góral, K., et al.: J. Phys. B **37**, 3457 (2004)
78. Vardi, A., Yurovsky, V.A., Anglin, J.R.: Phys. Rev. A **64**, 063611 (2001)
79. Liu, J., Liu, B., Fu, L.B.: Phys. Rev. A **78**, 013618 (2008)
80. Haque, M., Stoof, H.T.C.: Phys. Rev. A **71**, 063603 (2005)
81. Li, J., Ye, D.F., Ma, C., Fu, L.B., Liu, J.: Phys. Rev. A **79**, 025602 (2009)
82. Santos, G., Tonel, A., Foerster, A., Links, J.: Phys. Rev. A **73**, 023609 (2006)
83. Skornyakov, G., et al.: Zh. Eksp. Teor. Fiz. **31**, 775 (1956); Petrov, D.S., Salomon, C., Shlyapnikov, G.V.: Phys. Rev. Lett. **93**, 090404 (2004); Jensen, L.M., Makela, H., Pethick, C.J. Phys. Rev. A **75**, 033606 (2007)
84. Itin, A.P., Watanabe, S.: Phys. Rev. E **76**, 026218 (2007)
85. Tikhonenkov, I., Pazy, E., Band, Y.B., Fleischhauer, M., Vardi, A.: Phys. Rev. A **73**, 043605 (2006)
86. Liu, J., Fu, L.B., Liu, B., Wu, B.: New J. Phys. **10**, 123018 (2008)
87. Duine, R.A., Stoof, H.T.C.: Phys. Rep. **396**, 115 (2004)
88. Chen, Q., Stajic, J., Tan, S.N., Levin, K.: Phys. Rep. **412**, 1 (2005)

89. Inouye, S., Andrews, M.R., Stenger, J., Miesner, H.J., Stamper-Kurn, D.M., Ketterle, W.: Nature **392**, 151 (1998)
90. Timmermans, E., Furuya, K., Milonni, P.W., Kerman, A.K.: Phys. Lett. A **285**, 228 (2001); Holland, M., Kokkelmans, S.J.J.M.F., Chiofalo, M.L., Walser, R.: Phys. Rev. Lett. **87**, 120406 (2001); Ohashi, Y., Griffin, A. Phys. Rev. Lett. **89**, 130402 (2002)
91. Regal, C.A., Greiner, M., Jin, D.S.: Phys. Rev. Lett. **92**, 040403 (2004)
92. Bartenstein, M., Altmeyer, A., Riedl, S., Jochim, S., Chin, C., Hecker, J.D., Grimm, R.: Phys. Rev. Lett. **92**, 120401 (2004)
93. Zwierlein, M.W., Stan, C.A., Schunck, C.H., Raupach, S.M., Kerman, A.J., Ketterle, W.: Phys. Rev. Lett. **92**, 120403 (2004)
94. Regal, C.A., Ticknor, C., Bohn, J.L., Jin, D.S.: Nature **424**, 47 (2003)
95. Strecker, K.E., Partridge, G.B., Hulet, R.G.: Phys. Rev. Lett. **91**, 080406 (2003)
96. Cubizolles, J., Bourdel, T., Kokkelmans, S.J.J.M.F., Shlyapnikov, G.V., Salomon, C.: Phys. Rev. Lett. **91**, 240401 (2003)
97. Pazy, E., Tikhonenkov, I., Band, Y.B., Fleischhauer, M., Vardi, A.: Phys. Rev. Lett. **95**, 170403 (2005)
98. Altman, E., Vishwanath, A.: Phys. Rev. Lett. **95**, 110404 (2005)
99. Hodby, E., Thompson, S.T., Regal, C.A., Greiner, M., Wilson, A.C., Jin, D.S., Cornell, E.A., Wieman, C.E.: Phys. Rev. Lett. **94**, 120402 (2005)
100. Williams, J.E., Nygaard, N., Clark, C.W.: New J. Phys. **6**, 123 (2004); Williams, J.E., Nygaard, N., Clark, C.W.: New J. Phys. **8**, 150 (2006); Watabe, S., Nikuni, T.: Phys. Rev. A **77**, 013616 (2008)
101. Javanainen, J., Kotrun, M., Zheng, Y., Carmichael, A., Shrestha, U., Meinel, P.J., Mackie, M., Dannenberg, O., Suominen, K.A.: Phys. Rev. Lett. **92**, 200402 (2004)
102. Barankov, R.A., Levitov, L.S.: Phys. Rev. Lett. **93**, 130403 (2004); Andreev, A.V., Gurarie, V., Radzihovsky, L.: Phys. Rev. Lett. **93**, 130402 (2004); Dukelsky, J., Dussel, G.G., Esebbag, C., Pittel, S. Phys. Rev. Lett. **93**, 050403 (2004)
103. Miyakawa, T., Meystre, P.: Phys. Rev. A **71**, 033624 (2005)
104. Yi, W., Duan, L.M.: Phys. Rev. A **73**, 063607 (2006)
105. Kokkelmans, S.J.J.M.F., Milstein, J.N., Chiofalo, M.L., Walser R., Holland, M.J.: Phys. Rev. A **65**, 053617 (2002)
106. Chen, Q., Levin, K.: Phys. Rev. Lett. **95**, 260406 (2005)
107. Truscott, A.G., Strecker, K.E., McAlexander, W.I., Partridge, G.B., Hulet, R.G.: Science **291**, 2570 (2001); Hulet, R.G., Strecker, K.E., Truscott, A.G., Partridge, G.B.: Proceedings of XV International Conference on Laser Spectroscopy. World Scientific, Singapore (2002); Regal, C.A., Greiner, M., Jin, D.S.: Phys. Rev. Lett. **92**, 040403 (2004)
108. Diener, R.B., Ho, T.L.: arXiv: cond-mat/0405174 (2004)
109. Anglin, J.R., Vardi, A.: Phys. Rev. A **64**, 013605 (2001)
110. Yaffe, L.G.: Rev. Mod. Phys. **54**, 407 (1982)
111. Bergmann, K., Theuer, H., Shore, B.W.: Rev. Mod. Phys. **70**, 1003 (1998)
112. Brif, C., Chakrabarti, R., Rabitz, H.: New J. Phys. **12**, 075008 (2010); Ladd, T.D., Jelezko, F., Laflamme, R., Nakamura, Y., Monroe, C., Brien, J.L.O.: Nature(London) **464**, 45 (2010)
113. Rice, S.A., Zhao, M.: Optical Control of Molecular Dynamics. Wiley (2000)
114. Hänsch, T.W.: Rev. Mod. Phys. **78**, 1297 (2006); Hänsch, H., Biennier, L., Sims, I.R., Georgievskii, Y., Klippenstein, S.J., Smith, I.W.M.: Science **317**, 102 (2007)
115. Daems, D., Ruschhaupt, A., Sugny, D., Guérin, S.: Phys. Rev. Lett. **111**, 050404 (2013)
116. Schlosshauer, M.: Rev. Mod. Phys. **76**, 1267 (2004)
117. Hollenberg, L.C.L.: Nat. Phys. **8**, 113 (2012)
118. Král, P., Thanopulos, I., Shapiro, M.: Rev. Mod. Phys. **79**, 53 (2007)
119. Levitt, M.H.: Prog. Nucl. Magn. Reson. Spectrosc. **18**, 61 (1986)
120. Riebe, M., Monz, T., Villar, A.S., Schindler, P., Chwalla, M., Hennrich, M., Blatt, R.: Nat. Phys. **4**, 839 (2008); Monz, T., Kim, K., Hansel, W., Riebe, M., Villar, A.S., Schindler, P., Chwalla, M., Hennrich, M., Blatt, R.: Phys. Rev. Lett. **102**, 040501 (2009)

121. Piltz, C., Scharfenberger, B., Khromova, A., Varon, A.F., Wunderlich, C.: Phys. Rev. Lett. **110**, 200501 (2013); Hill, C.D.: Phys. Rev. Lett. **98**, 180501 (2007)
122. Genov, G.T., Vitanov, N.V.: Phys. Rev. Lett. **110**, 133002 (2013)
123. Dridi, G., Guérin, S., Hakobyan, V., Jauslin, H.R., Eleuch, H.: Phys. Rev. A **80**, 043408 (2009)
124. Ibáñez, S., Chen, X., Torrontegui, E., Muga, J.G., Ruschhaupt, A.: Phys. Rev. Lett. **109**, 100403 (2012); Torrontegui, E., Ibáñez, S.S., Martłnez-Garaot, S., Modugno, M., del Campo, A., Guéry-Odelin, D., Ruschhaupt, A., Chen, X., Muga, J.G.: Adv. At. Mol. Opt. Phys. **62**, 117 (2013)
125. Bason, M.G., Viteau, M., Malossi, N., Huillery, P., Arimondo, E., Ciampini, D., Fazio, R., Giovannetti, V., Mannella, R., Morsch, O.: Nat. Phys. **8**, 147 (2012); Demirplak, M., Rice, S.A.: J. Phys. Chem. A **107**, 9937 (2003); Berry, M.V.: J. Phys. A **42**, 365303 (2009)
126. Torosov, B.T., Guérin, S., Vitanov, N.V.: Phys. Rev. Lett. **106**, 233001 (2011)
127. Vitanov, N.V.: Phys. Rev. A **84**, 065404 (2011); Kyoseva, E., Vitanov, N.V.: Phys. Rev. A **88**, 063410 (2013); Torosov B.T., Vitanov, N.V.: Phys. Rev. A **87**, 043418 (2013); Genov, G.T., Schraft, D., Halfmann, T., Vitanov, N.V. Phys. Rev. Lett. **113**, 043001 (2014)
128. Schraft, D., Halfmann, T., Genov, G.T., Vitanov, N.V.: Phys. Rev. A. **88**, 063406 (2013)
129. Trimborn, F., Witthaut, D., Kegel, V., Korsch, H.J.: New J. Phys. **12**, 053010 (2010)
130. Dou, F.Q., Cao, H., Liu, J., Fu, L.B.: Phys. Rev. A **93**, 043419 (2016)
131. Fu, L.B., Liu, J.: Phys. Rev. A **74**, 063614 (2006)
132. Liu, J., Wu, B., Fu, L.B., Diener, R.B., Niu, Q.: Phys. Rev. B **65**, 224401 (2002)
133. Albiez, M., Gati, R., Fölling, J., Hunsmann, S., Cristiani, M., Oberthaler, M.K.: Phys. Rev. Lett. **95**, 010402 (2005)
134. Khomeriki, R.: Phys. Rev. A **82**, 013839 (2010)

Index

J. Liu et al., *Nonlinear Adiabatic Evolution of Quantum Systems*,
https://doi.org/10.1007/978-981-13-2643-1

MIX
Papier aus verantwortungsvollen Quellen
Paper from responsible sources
FSC® C105338

If you have any concerns about our products,
you can contact us on
ProductSafety@springernature.com

In case Publisher is established outside the EU,
the EU authorized representative is:
Springer Nature Customer Service Center GmbH
Europaplatz 3, 69115 Heidelberg, Germany

Printed by Libri Plureos GmbH
in Hamburg, Germany